D1068534

Beginnings of
Brazilian Science

BEGINNINGS OF Brazilian Science

Oswaldo Cruz, Medical Research and Policy, 1890–1920

NANCY STEPAN

Science History Publications
New York
1976

Science History Publications
a division of
Neale Watson Academic Publications, Inc.
156 Fifth Avenue, New York, N.Y. 10010

Library of Congress Cataloging in Publication Data

Stepan, Nancy.
Beginnings of Brazilian science.

Bibliography: p.187
1. Medicine—Brazil—History. 2. Cruz, Oswaldo Gonçalves, 1872-1917. 3. Underdeveloped areas—Science. I. Title. [DNLM: 1. Research—History—Brazil. 2. Science—History—Brazil. Q180.B7 S827b]
R482.B8S74 610′.981 75-7796
ISBN 0-88202-032-3

Designed and manufactured in U.S.A.

To my parents, Duncan and Erica,
and my husband, Alfred

Contents

	Preface	ix
1	Introduction: Statement of the Problem	1
	References	10
2	Science in Brazil Before 1900: The "Colonial" Tradition	13
	Colonial Science in Spanish and Portuguese America	16
	Science in Nineteenth Century Brazil	23
	Science in the Old Republic	29
	The Setting in 1900	36
	References	40
3	Medicine in Brazil: The Nineteenth Century Background	47
	Medicine in the Nineteenth Century	50
	Medicine in the 1900s	56
	References	60
4	Epidemic Disease and the Growth of Science: The Serum Therapy Institute of Rio de Janeiro	66
	Oswaldo Cruz and His Training	69
	Early Years at the Serum Therapy Institute	73
	Models in Science: The Pasteur Institute	75
	References	80
5	From Serum Therapy to Research: Science and Politics in Brazil, 1903-1908	84
	Cruz and the Yellow Fever Campaign	85
	The Congressional Debate on Science	91
	The Founding of the Oswaldo Cruz Institute	97
	References	101
6	The Survival of Science in a Developing Country: Students, Clients and Research	105
	Staffing of the Institute	106
	Clients and the Uses of Science	112
	Research Science at the Institute	115
	"National" Science and "International Science"	120
	Conclusion	125
	References	128

7 The Bacteriological Institute of São Paulo, 1892–1914: The Role of Applied Science 134
São Paulo and Public Health 136
The Bacteriological Institute: Its Scientific Work 138
Research, Students, and Applied Science in the Bacteriological Institute 144
The Decline of the Institute 150
References 153

8 Science in a Developing Country: Some Policy Issues 157
Introduction 157
National Capabilities in Research and Applied Science 157
National Capabilities in Technology 166
The Meaning of "National" Science 170
The Setting for Science: Universities and Research Institutes 176
Conclusion 180
References 181

Selected Bibliography 187

Plates 209

Index 219

Preface

This book is based on research carried out during several trips to Brazil, where my main interest was in studying the founding and maintenance of institutions of science. I was fortunate that the archives of the Oswaldo Cruz Institute (Instituto Oswaldo Cruz), one of the most successful institutions of science in the early twentieth century in Brazil, and, indeed, in Latin America, were made accessible to me. I wish to thank the members of the department of administration at the Oswaldo Cruz Institute for making it possible for me to examine various archives, including administrative records and a collection of uncatalogued manuscripts housed in the institute's museum. These materials are described in the bibliography. The library staff were most courteous in aiding my search for books and articles on the history of the institute, and in providing me with copies of photographs of the institute's early years. Some of these photographs are reproduced, with acknowledgment, in this book. I particularly want to thank Senhora Emilia Bustamente, chief librarian while I was undertaking research, for her permission to use in Fig. 1 material she prepared showing the numerical growth in the publications and staff size of the institute.

In Brazil I was also able to consult publications and archival material relating to other institutions of the period, such as the Bacteriological Institute (Instituto Bacteriológico) and the Butantã Institute (Instituto Butantã), both in São Paulo. Professor Francisco Bruno Lobo, who is making a study of the history of the medical school in Rio de Janeiro, very kindly allowed me to read as yet unpublished, mimeographed copies of the official reports of the medical faculty.

I was also fortunate in being able to talk to a number of people involved either at first- or second-hand with the scientific movement that led to the creation of the first internationally recognized school of medical research in Brazil. Senhora Henrique da Rocha Lima, widow of one of Oswaldo Cruz' most important colleagues, received me into her home to permit me to copy letters written by Cruz to Rocha Lima in the formative period of the Oswaldo Cruz Institute's history. These letters have subsequently been published, as described in the bibliography. Drs. Olympio da Fonseca and Artur Moses, both former members of the institute and both closely associated with the development of the biomedical sciences in Brazil, were generous with their time in answering my questions and providing me with valuable insights and information. Dr. Oswaldo Cruz Filho and his family extended their friendship to me and allowed me to study their extensive files on the life of Oswaldo Cruz.

Two comments about the spelling and title of this book are in order before I make further acknowledgments. Portuguese spelling has undergone several changes in the course of the twentieth century. In all but a few cases I have chosen to use modern Portuguese orthography in the text (e.g., Adolfo), but have retained the original spelling of names and words in the books and articles cited in the references (e.g., Adolpho), in the belief that this would be helpful to other historians. This explains apparent inconsistencies in spelling in the text and references; sometimes a name will even be spelled two ways in a single reference, depending on the dates of the publications cited. Throughout the book I have standardized the use of accents. One exception to the use of modern spelling in the text is Oswaldo Cruz' name. Modern orthography would dictate Osvaldo, but, since the institute is still called the Oswaldo Cruz Institute, Oswaldo seemed the most appropriate spelling in my book.

Concerning the title of the book, it is obvious from my long discussion of the history of the sciences in Brazil before 1900 that there was much science in Brazil before Oswaldo Cruz and the institute he created. However, my emphasis is on research science as an organized, institutional endeavor, and in this respect the Oswaldo Cruz Institute represents the beginnings of Brazilian science.

In the United States and Brazil several scholars helped with comments, criticisms, and suggestions during the writing of the book. I approached my task as a historian of science rather than as a historian of Brazil. I am, therefore, greatly indebted to specialists in the history of Brazil for saving me from errors of fact and interpretation. I especially want to thank Professors Richard M. Morse, of Yale University, Thomas E. Skidmore, of the University of Wisconsin, and Joseph L. Love, of the University of Illinois at Urbana-Champaign, for their close reading of the manuscript and their suggestions for revisions. It goes without saying that any errors that remain are my responsibility. Professors Alberto Guerreiro Ramos, of the University of Southern California, and Simão Mathias, of the Instituto de Química, University of São Paulo, Brazil, both gave me the benefit of their knowledge of the history of science in Brazil.

Professor John G. Burke, of the University of California at Los Angeles, has been an invaluable teacher, critic, and friend since the inception of this book. Professor Albert O. Hirschman, of the Institute for Advanced Study, Princeton, read a portion of an earlier version of this book and encouraged me to continue the project. I have also learned from Professor Richard R. Nelson, whose knowledge of the science policy field was very useful to me when my work led me into this area.

Part of the research for this book was supported by a National Science Foundation Traineeship in the history of science at the University of California at Los Angeles, and by a National Institutes of Health Fellowship. A leave of absence from the Department of History at the University of Massachusetts at Amherst, and residence at Yale University as a Visiting Fellow in the Department of the History of Science and Medicine in the Fall of 1974, gave me the time necessary to write the last chapter of the book. At Yale, Professor Derek de Solla Price was an invaluable colleague whose ideas on science and science policy were always stimulating.

Since it seems the custom to name last rather than first the person to whom one owes the most in a project such as writing a book, I come finally to my husband Alfred. As a political scientist with a special interest in Brazil, his knowledge was always very useful. But, more important, he never failed to give me all the support he could in the late nights of writing, and share with me the pleasures of children and the pleasures and pains of work. To him, then, go my deepest thanks.

NANCY STEPAN

Amherst, Massachusetts
and
New Haven, Connecticut

1

Introduction: Statement of the Problem

The subject of this book is how, when, and why western science began to become established in Brazil. Since Brazil is a developing country not especially known for its contributions to science, it might be asked why a historian of science has chosen to study this particular problem.

Traditionally, of course, historians have shown a strong interest in the growth of science in what are now the industrialized and developed countries of the world. Many aspects of this development have been studied with some thoroughness—the connection between the intellectual heritage of Greek and medieval science and the burst of intellectual activity in the sixteenth and seventeenth centuries known as the scientific revolution, the relation between Protestantism and capitalism and the growth of science, and the impact of industrialization on scientific activities.[1] But relatively few historians have concerned themselves with the question of how and why western science has spread from the countries of Europe to other areas of the world, especially developing countries.[2] This neglect of what is surely a central problem in the history of science is somewhat surprising, given the importance that science and technology are believed to have for the process of economic development, and the concern that has been expressed about the gap between the industrialized, scientific, rich nations, and the industrializing, semi-scientific, poor nations of the world.[3]

A study, therefore, of the growth of science in one of these latter countries seemed justified as the long-overdue extension of the historian's traditional interest in the history of science to the new field of developing countries. It would also serve to alert those involved in science policy, as well as historians, to a set of factors that figure prominently in the emergence of science in developing countries, factors that may have been overlooked through an exclusive focusing on problems of science in developed areas. We need to know much more about the complex political, economic, and cultural supports required for the successful pursuit of science in different periods of time. An analysis of the impact of European science in Brazil, and the evolution and adaptation of scientific institutions in the Brazilian context, would raise many interesting questions about traditional obstacles to science, the criteria for scientific development, the role of planning and the government, and the desired balance between theoretical and applied science in a nation with limited resources.

The studies of the historical rise of science in Europe and other industrialized countries have given some indication of the range of factors involved in the origin and maintenance of sophisticated and highly productive scientific communities. Following the emergence of science as an intellectual activity in the seventeenth century, science in Europe was spurred to further development, and scientific organization given new shape, by rapid economic and industrial growth starting in the late eighteenth century. After 1870, as industrialization began to create a demand for trained scientists and technologists, science came to be supported by industry, as a source of profit, and by national governments, for the economic and technological benefits it brought. World War II saw the advent of large-scale science, involving the financial and intellectual resources of a large sector of the industrialized countries.[4]

Today science is increasingly carried out not by individuals but by groups of scientists working together on common research problems, in research institutes, industrial laboratories, university science departments, and government science agencies. The values and ideas of scientists are transmitted to new generations of science students in these institutions. Giving support to the scientific enterprise are universities and secondary schools, whose curricula include the study of scientific subjects. Additional support is given by specialized scientific libraries and factories producing technical equipment. Opportunities for employment in science have risen, encouraging the pursuit of scientific careers on a full-time basis. National and international networks of communication between scientists have been established, putting scientists in different communities and in different countries in touch with each other. Publication in scientific journals, and membership in scientific societies, give recognition to the achievements of scientists, while at the same time setting up standards of excellence and mechanisms for quality control. The "republic of science", described by Polanyi as a semi-autonomous republic with its own scientific authority, has come into being.[5]

Developing countries such as Brazil have been at the periphery of this process for a number of reasons. One reason has been their initially slow and later rapid but often "dependent" economic and industrial growth, which has limited the incentives for scientific growth. Scientists in Brazil have been isolated from the "republic" because of their inability to receive adequate training in science at home or to find it abroad in sufficient numbers, because of poor opportunities for employment, and because of the lack of supporting institutions for science, such as libraries and schools reflecting the values of science. More than this, a country such as Brazil has historically presented a series of obstacles to the development of science, the full nature of which cannot be properly indicated by the study of the characteristics of advanced scientific communities of the world, or by noting the absence of or the need to create these characteristics in the developing country. Religious dogmas, traditional patterns of education, or traditional values held by the elites, help shape a culture in particular ways and in turn may act to inhibit the establishment of ongoing communities of science. For this reason, the study of the advanced scientific community cannot tell us a great deal about how a developing country begins the process of transformation of its semi-scientific community into a fully successful, scientific one. Many of the characteristics of the advanced scientific community, for example, are effects rather than causes of the advance of science. We still know little about which factors are essential for the growth of science in countries that historically did not participate in the scientific or industrial revolutions.

Indeed, the tacit assumption that the path of science to be followed in developing countries will be similar to that already taken by industrial countries, and that success in science depends upon identifying and eliminating obstacles to science as understood by the study of science in countries "further ahead" on the road towards scientific independence, needs to be reexamined. It may be that the obstacles to the growth of science in developing countries are qualitatively different from those in developed countries, and of a kind not readily apparent to traditional historians of science.

This last point is in fact the burden of the argument made by a group of social scientists and economists, mainly from the developing world, and commonly referred to as "dependency theorists." They argue that developing countries, coming late to the industrialization process and dependent upon the industrial world for capital and scientific and technological knowledge, encounter a series of obstacles to the growth of science that analysts of European and North American science have not even considered. They suggest, convincingly in the broad range, that the history of the industrialization process in developing countries is such as to make the growth of science and technology inherently different from that in the industrial world. Their work puts the history of science and technology in countries like Brazil in a new perspective. In particular, dependency theorists stress the importance of creating autonomous, indigenous capabilities in science and technology in peripheral countries.[6] In this respect, my own study of science in Brazil contributes to this debate, because its main theme is the difficulties related to the creation of indigenous strengths in science in a developing country.

When I began this book, I asked myself the general question of when and how science began to be carried out in Brazil. Brazil was a country I already knew and it seemed an interesting country to study for a number of reasons. It shared with most developing countries the fact of being relatively untouched by the scientific revolution of the sixteenth and seventeenth centuries, and the industrial revolution of the nineteenth century. Unlike Japan, which had successfully industrialized and acquired scientific capabilities in the late nineteenth and early twentieth centuries, Brazil possessed classic characteristics of underdevelopment, such as low literacy rates, high unemployment, and poorly developed communications. At the same time, and unlike many of the other developing countries, Brazil was a large country, with a long history of political independence, a varied indigenous culture, and a complex set of political and intellectual institutions. In addition, and again unlike many other developing countries, Brazil has had a rich and more or less continuous contact with European thought since Brazil was discovered by Cabral in 1500. Brazil has in fact been a "consumer" of European ideas, European science, and European technology ever since the beginning of its history. Yet until recently scientific activity had been only sporadic.

As my research into science in Brazil got underway, it became clear that rather than investigate the vast but negative question of why science failed to get established before the twentieth century, a better strategy was to examine some point in Brazil's history where a break occurred in the traditional indifference to science, and where science began to be pursued with some degree of success. By comparing this period with earlier ones, it seemed that some of the factors involved in the successful establishment of science might become clear. The creation of indigenous capabilities in science thus became the chief topic of enquiry.

A reading of the literature on Brazil suggested that the period immediately after 1900 represented such a break, especially as exemplified in a particular scientific institution known eventually as the Instituto Oswaldo Cruz (the Oswaldo Cruz Institute). Oswaldo Cruz is well-known to Brazilians and historians of Brazil as the "sanitizer" of Rio de Janeiro who, as director of public health between 1903 and 1909, rid the capital of yellow fever and bubonic plague. His directorship of a scientific institution is also known but its overall significance for science in Brazil is less well appreciated.[7] In fact, the Oswaldo Cruz Institute was the first research institute, properly speaking, in Brazil's history, the first to make scientific contributions over a sustained period of time, and the first to give Brazil a reputation abroad for science.[8] The fact that the institute represented a success for Brazil seemed to guarantee that its history would be available for study. Failed institutions do not always keep their records for future historians to examine. The surmise that the Oswaldo Cruz Institute represented an important first step in the eventual transformation of Brazil's scientific capabilities proved correct. Not only did the institute possess rich manuscript and published sources concerning its origin and early history, to which I was given access, but its existence did signal a genuinely radical departure from the earlier tradition of science established in nineteenth century Brazil.[9] The institute was founded in 1900 outside the capital of Rio de Janeiro as a small municipal laboratory where vaccines against the bubonic plague could be produced. From these modest beginnings the institute expanded in a short time into a center of experimental medicine, where a "school" of Brazilian microbiologists and protozoologists carried out scientific work of a high quality for many years. By 1908, in fact, the institute was sufficiently well-established to move into modern laboratories and attract to Brazil foreign scientists interested in tropical medicine. At the institute researchers found the supports necessary for serious scientific work.

The institute's rise to fame was closely associated with the public health campaign against yellow fever in the capital city of Rio de Janeiro, which began in 1903. From this date on, the institute was responsible for a large number of programs in practical hygiene. In addition, however, the Oswaldo Cruz Institute made many contributions to medical research. In 1909, indeed, the institute made medical history by the announcement of the discovery of a hitherto completely unrecognized human disease that ravaged populations over large areas of South and Central America, as well as some southern areas of the United States.[10] This disease was American sleeping-sickness (*Trypanosomiasis americana*), commonly called "Chagas' disease" after its discoverer. The description and analysis of the symptoms, etiology, and pathology of this disease became one of the chief research activities of the institute.[11]

In that same year, 1909, the institute founded its own journal, the *Memorias do Instituto Oswaldo Cruz,* long regarded by medical scientists as one of the few significant research journals coming out of Latin America, and organized the first modern courses in microbiology in the country. In 1907, even prior to the discovery of Chagas' disease and the series of studies it precipitated, the institute received the gold medal at the International Conference of Hygiene at Berlin for its contributions to the hygiene sciences. Partly as a result of this international recognition, the Brazilian government named the institute after its director, Oswaldo Cruz, and consolidated its financial and administrative status. In 1913, Cruz was made a member of the Brazilian Academy of Letters. In 1917, at the age of only forty-five, Cruz died of renal disease. His successor, Carlos Chagas, had meanwhile acquired a national and international reputation matching Cruz' own and under Chagas the institute was able to expand its departments further. The continued growth of the institute, as measured by the number of publications by institute scientists and the number of scientists employed by the institute, is shown in Fig. 1. A tradition of biomedical research was established that provided the foundation for the more or less continuous development of the biomedical sciences in Brazil from the first decades of the century to the present, even though this development was often slow and uneven, and the understanding of the need for long-range research and a commitment to public health slow to develop.

The history of the Oswaldo Cruz Institute is also important from another perspective. Its founding and growth coincided with what is known as "la belle époque" in Brazilian history, the period immediately after 1900 when Brazilians sought to define Brazilian culture in new terms, and to make Brazil an active partner in world civilization. Wider contacts between Brazilians and Europeans, brought about by travel in Europe, had made Brazilians conscious of the gap between Europe and Brazil, and between Brazilian potentiality and Brazilian reality. There was felt both a desire to emulate European, particularly French, culture, and to lessen dependence on European values and assert a national identity. In the period we see much optimism about the possibility of material progress, on the one hand, and pessimism about such progress on the other. We see a growing pride in Brazil, of which the rebuilding of the federal capital of Rio de Janeiro was one manifestation. Yet the period also saw the publication of Euclides da Cunha's masterpiece, *Os Sertões* (translated into English as *Rebellion in the Backlands*), which, in recounting the story of an armed rebellion in the northeast, contrasted in stark terms life in the cities with life in the vast hinterland, where Brazilians lived in wretchedness and poverty.

The Oswaldo Cruz Institute contributed to both the optimism and the pessimism of the period. The sanitation of Rio de Janeiro and the founding of the first scientific research establishment were sources of immense satisfaction to Brazilians and a sign of the capacity of Brazil to contribute to world civilization. Yet medicine also proved a vehicle by which the terribly diseased state of the national health was revealed. The study of the history of the Oswaldo Cruz Institute in Brazil, therefore, makes a contribution to a neglected phase of Brazilian history. It also allows the historian to explore the themes of dependency and nationalism in the creation of national communities of science.

The success of the Oswaldo Cruz Institute in drawing together a large number of Brazilian-trained scientists in a single place to focus on a series of related research problems, in originating scientific work of genuine significance, in securing sufficient financial and administrative autonomy to weather political change and Cruz' own death, was especially remarkable given the fact that the crucial years of the institute occurred quite independently of any national plan for the advancement of science. As a case history, the story of the institute promised therefore to be extremely rewarding from the point of view of how and why science develops to a new level of effectiveness in a developing country still at the early stages of economic growth and industrialization. The institute's solutions to many of the problems concerning the supports necessary for science in a country with a small scientific tradition provided me with a basic insight concerning the central role of institutions, an insight I develop throughout the book, ending with the study's policy implications, which I discuss in the last chapter. The examination of the tradition of science established by the end of the nineteenth century, the comparative successes and failure of institutes being founded contemporaneously with that of the Oswaldo Cruz Institute, and the long-range impact the Institute had on the scientific tradition within Brazil, provided the overall framework for this book.

The history of the biomedical sciences in general, and the Oswaldo Cruz Institute in particular, also suggested new definitions of what "successful" science might mean in a developing country. By the criteria commonly used in the industrial world, such as the number of scientific publications produced by a country, the number of scientific discoveries made, or the number of Nobel prizes won, neither Brazil nor any of the other countries of the developing world are yet successful in science.[12] But such criteria ignore the contributions that scientific work makes to a country itself, even when these contributions, measured on a world scale, are small. The history of the biomedical sciences in Brazil between 1900 and 1920 suggests the appropriateness of rather different criteria. Success would be measured by the creation of stable and productive institutions of fundamental and applied research. Success would mean the ability of an institution to survive over time and to diversify its staff and range of activities. Success would refer to the continued ability of an institution to recruit scientists, and to contribute to the absorption of national scientific and technical manpower. Success would refer to the ability of an institution to increase support for science. It would be measured in terms of the institutions's influence on other institutions of science within the country. In terms of output, success would refer to the ability of an institution to produce science that either serves local needs, or results in the understanding of national scientific problems (for example, tropical disease), or profits from local factor endowments, rather than depending on the international world of science for the definition and choice of subjects for study. At the same time, successful science in a developing country must not be cut off from that international world of science, but must find an "audience" for its work and thus contribute to the general growth of knowledge.

These criteria of "success" are explored as the book unfolds. In Chapters 2 and 3 I describe in general terms how Latin America first came into contact with European science, and more specifically, how the scientific and medical community of Brazil evolved in the eighteenth and nineteenth centuries. The typical institutions, bureaucratic relations and values of science in Brazil are described, and the characteristic weaknesses of the scientific tradition before 1900 evaluated. These chapters indicate the range of personal, political, and institutional mechanisms that had to be created before a nucleus of experimental science could be successfully created in Brazil.

The obstacles to science in the period during which Oswaldo Cruz worked were such, in fact, as to involve Cruz in a complicated process of institution-building and entrepreneurship of a type in some ways different from that marking the development of scientific nuclei in the developed world. This process is explored in Chapters 4, 5, and 6. In Chapter 4, the immediate medical and political reasons leading to the establishment of the institute as a serum therapy laboratory in 1900 are described, and Oswaldo Cruz' first association with the laboratory discussed. Chapter 5 examines how the institute grew into a "Pasteur Institute" in Brazil between 1903 and 1909, during Cruz' tenure of the post of director of public health. The role of politics and the entrapreneur in science are analyzed in detail. In Chapter 6 the relation between pure and applied research as it emerged in the institute, the function of students, and the need for supporting institutions are examined. The chapter concludes with a more general evaluation of the Oswaldo Cruz Institute's impact on the tradition of science and scientific administration in Brazil. Obviously, science was not transformed overnight in Brazil, and the limitations of the institute are discussed.

Chapter 7 turns from the Oswaldo Cruz Institute to examine the history of a similar institution of science being founded at roughly the same time, the Bacteriological Institute of São Paulo, directed by the distinguished Swiss-born biologist, Adolfo Lutz. While this latter institution was one of the keys to the emergence of the state of São Paulo as a leader in the public health movement in Brazil, the institute nonetheless failed to develop into a center of experimental medicine.

The reasons for this, and the bearing it has for understanding the advancement of science in Brazil in this period, provide the focus of the chapter.

In Chapters 2-7 I write as a historian of science with a story to tell. In Chapter 8 I write as a sociologist of science and from the perspective of science policy. I try to sum up the broader implications of the history of the biomedical sciences in Brazil between 1890 and 1920. The history of the Oswaldo Cruz Institute belongs to a period when industrialization was just beginning in Brazil. I am very much aware of the present-day difficulties standing in the way of the creation of scientific and technological capabilities in developing countries. I make no claim that the Oswaldo Cruz Institute led to an immediate or startling change in science in Brazil. Throughout the book, however, I do argue that the history of the Oswaldo Cruz Institute provides an example of some very interesting solutions to the problems attendant upon creating independent and relatively autonomous abilities in science. Its history allows me, in Chapter 8, to discuss such questions as the criteria a developing country must use in choosing between different kinds of science to support and the best way to achieve useful applications of science. That is, the history of the Oswaldo Cruz Institute allows me to address some of the critical questions of science in a developing country. The fact is that there are very few examples of successful science in Latin America to date, and that one of the most interesting is found in the biomedical sciences in Brazil. For this reason I believe that the history of the Oswaldo Cruz Institute is of general significance.

References

[1]Some basic works in this field are Frederick B. Artz, *The Development of Technical Education in France, 1500–1850* (Cambridge, Massachusetts: Cleveland Society for the History of Technology, 1966); George Basalla (*ed.*), *The Rise of Modern Science: Internal or External Factors?* (Lexington, Massachusetts: D. C. Heath and Company, 1968), Joseph Ben-David, *The Scientist's Role in Society. A Comparative Study* (Englewood Cliffs, New Jersey: Prentice-Hall, Inc., Foundations of Modern Sociology Series, 1971); Edwin A. Burtt, *The Metaphysical Foundations of Modern Physical Science; A Historical and Critical Essay* (London: Routledge and Kegan Paul, 1949); Marshall Clagett, *The Science of Mechanics in the Middle Ages* (Madison: University of Wisconsin Press, 1959); A. C. Crombie, *Robert Grosseteste and the Origins of Experimental Science, 1100–1700* (Oxford: Clarendon Press, 1953); A. C. Crombie (*ed.*), *Scientific Change; Historical Studies in the Intellectual, Social and Technical Conditions for Scientific and Technical Invention, from Antiquity to the Present* (New York: Basic Books, 1963); Alexandre Koyré, *From the*

Closed World to the Infinite Universe (Baltimore: Johns Hopkins Press, 1957); Robert K. Merton, *Science, Technology and Society in Seventeenth Century England* (New York: Harper and Row, 1970); John H. Randall, *The School of Padua and the Emergence of Modern Science* (Padova: Ed. Antenore, 1961); Lynn Thorndike, *A History of Magic and Experimental Science*, 8 vols. (New York: Columbia University Press, 1923–58); Richard S. Westfall, *Science and Religion in Seventeenth-Century England* (New Haven: Yale University Press, 1958); W.P.D. Wightman, *Science and the Renaissance* (Edinburgh: Oliver and Boyd, 1962).

[2]A major exception to this is, of course, the case of the United States. Here the absorption and subsequent evolution of European science in the American context has received its share of scholarly studies. Again, the literature is enormous, but among other books see John C. Burnham, *Science in America; Historical Selections* (New York: Holt, Rinehart and Winston, Inc., 1971); George H. Daniels, *American Science in the Age of Jackson* (New York: Columbia University Press, 1968); A. Hunter Dupree, *Asa Gray, 1810–1888* (New York: Atheneum, 1968); Brooke Hindle, *The Pursuit of Science in Revolutionary America, 1735–1789* (Chapel Hill: University of North Carolina Press, 1956); Joseph F. Kett, *The Formation of the American Medical Profession: The Role of Institutions, 1780–1860* (New Haven: Yale University Press, 1968); Edward Lurie, *Louis Agassiz; A Life in Science* (Chicago: University of Chicago Press, 1960); Thomas G. Manning, *Government in Science; The U.S. Geological Survey, 1867–1894* (Lexington: University of Kentucky Press, 1967); Howard Smith Miller, *Dollars for Research; Science and its Patrons in Nineteenth-Century America* (Seattle: University of Washington Press, 1970); James L. Penick *et al.* (*eds.*), *The Politics of American Science, 1939 to the Present* (Chicago: Rand McNally, 1965); Carroll W. Pursell, Jr. (*ed.*), *Readings in Technology and American Life* (New York: Oxford University Press, 1969); Nathan Reingold (*ed.*), *Science in Nineteenth-Century America, A Documentary History* (New York: Hill and Wang, 1964); Richard Harrison Shryock, *Medicine in America, Historical Essays* (Baltimore, Maryland: The Johns Hopkins Press, 1966); Raymond P. Stearns, *Science in the British Colonies of America* (Urbana: University of Illinois Press, 1970); Dirk J. Struik, *Yankee Science in the Making* (Boston: Little, Brown, 1948); David D. Van Tassel and Michael G. Hall, (*eds.*), *Science and Society in the United States* (Homewood, Illinois: The Dorsey Press, 1966).

[3]While it is clear there is a positive relationship between science and industrial growth today, the relationship is recent and not entirely understood. See Chapter 8 for further discussion.

[4]See Joseph Ben-David, *op. cit.*, for a succinct analysis of these trends and a good introduction to the literature on the subject. The character of the modern scientific and research system of the industrialized countries is discussed at greater length in Chapter 8.

[5]Michael Polanyi, "The Republic of Science: Its Political and Economic Theory," in Edward A. Shils (*ed.*), *Criteria for Scientific Development: Public Policy and National Goals; A Selection of Articles from Minerva* (Cambridge, Massachusetts: The M.I.T. Press, 1968), pp. 1-20.

[6]For a discussion of the dependency literature and an analysis of the obstacles to science and technology, see Chapter 8.

[7]The Oswaldo Cruz Institute is mentioned, for instance, in Fernando de Azevedo, *As*

ciências no Brasil, 2 vols. (Rio de Janeiro: Edições Melhoramentos, 1955), Vol. 2, pp. 224-234.

[8]Recently M. Frota Moreira identified the Oswaldo Cruz Institute as representing the first stage in the creation of a research science tradition in Brazil. See M. Frota Moreira, "Prioridades e objetivas nacionais de desenvolvimento," in Heitor G. de Souza *et al.*, *Política científica* (São Paulo: Editôra Perspectiva, 1972), pp. 269-283.

[9]See the bibliography for a description of these sources.

[10]In 1913 Chagas received the Schaudinn prize for this discovery. The prize, named in honor of Fritz Schaudinn, the discoverer of the *Treponema pallidum* of syphilis, was awarded every four years to the scientist who had done the most to advance the field of protozoology. For the controversies about Chagas disease—its pathology, incidence, and relation to other diseases—see Chapter 6, section *Research Science at the Institute.*

[11]The various contributions and activities of the Oswaldo Cruz Institute are discussed fully in Chapters 4, 5, and 6.

[12]Ninety percent of the world's science, as measured by publications, rests in 14 countries (USA, UK, USSR, Germany, France, Japan, Canada, India, Italy, Australia, Switzerland, Czechoslovakia, Sweden, Netherlands). See Derek de Solla Price, "Measuring the Size of Science," *Proceedings of the Israel Academy of Sciences and Humanities* 4 (1969), p. 105, and Table I, p. 106. Two Nobel prizes have been awarded to scientists from the 20 countries of Latin America (both to Argentina, to B. Houssay in medicine in 1947, and L. F. Leloir in chemistry in 1970).

2

Science in Brazil Before 1900: The Colonial Tradition

In the long interval between the discovery of the New World and the present, few scientists have risen to prominence in the countries of Latin America. Few from this area of the world have contributed to the seminal ideas that dominate the work of generations of scientists and advance our knowledge of the natural world. Even at the middle and lower ranges of science, Latin American contributions have been small in number.

The lack of science in Latin America would be less interesting to the historian of science if it were the result of Latin America's isolation from European scientific thought. As we stated earlier, however, contact with European thought has been more or less continuous since the sixteenth century. Men trained in Europe brought ideas and information to the Spanish and Portuguese colonies in America, and explored the native flora and fauna of the New World. Some Latin Americans were also stimulated to participate in the process of scientific exploration of their native lands, and to begin to study science for its own sake. Yet despite contact and at least some interest in western science, only very slowly did productive communities of science develop, and in most countries of Latin America only very recently.

The absence of science in Latin American culture has been commented upon mainly by omission—omission of discussions of science and its role in history in the standard books on Latin America. Occasionally attention has been paid to some aspect of Latin American culture that sheds light on science.[1] Some Latin American historians have charted the course of science in individual countries but few of the accounts relate these developments adequately to the history of science elsewhere in the continent. Those best placed to evaluate the slow rise of science in Latin America and to place this rise in the perspective of the general spread of western science in the world, namely the historians of science, have shown little interest in the development of science outside the major scientific countries. As a consequence, the important questions of how and why science has spread and been adapted to cultures outside of Europe, and particularly questions about the evolution of science in "developing" areas, have been ignored.

One exception to the general lack of attention paid to an interesting problem in the history of science is the brief but suggestive article published by the historian, George Basalla, entitled "The Spread of Western Science."[2] Basalla asks how European science diffused from Italy, France, England, the Netherlands, Germany, Austria, and the Scandinavian countries to other areas of the world, including the United States, Japan, Canada, Australia, China, and Latin America. He proposes a model for understanding this process which divides the stages of growth into three. Stage one occurs when contact is made with a new country by western Europeans, through military conquest, colonization, missionary work, or commercial activity, or a combination of all or some of these factors. Whatever the method of contact, however, in stage one the new country serves primarily as a source of interesting data about plants, animals, minerals, and peoples, data which are sent back to Europe for absorption, classification, and analysis. Stage one science occurs as an extension of the general process of exploration, and as such the sciences associated with exploration dominate in this phase—natural history and anthropology, with cartography, topography, astronomy, and meteorology occasionally rising to importance.

As native scientists join in the enterprise of scientific exploration, indigenous institutions of science begin to be founded. There is also a widening of interests until finally most of the fields of science undertaken in the sponsoring or supporting European country are practiced in the colonial country. Basalla calls this second stage of science "colonial" because, although the numbers of people involved in science are increasing, nonetheless the scientific community is still dependent upon traditions in science lying outside the country. The practitioners of colonial science continue to be trained in Europe until the time when native schools are capable of providing scientific education at home. Few of the institutions founded are large enough to sustain a research program. Yet enough of an intellectual *milieu* is maintained to explain the rare emergence of genuinely original scientists. These men, as Basalla observes, become the heroes of colonial science. Benjamin Franklin is a classic example of the colonial scientists whose creativity and whose ties to Europe are widely recognized.

The term "colonial" science is used to describe a dependent tradition even when politically no colonial arrangement exists. Science in the United States, for example, was "colonial" for some time after independence. Eventually, however, native-born scientists begin to reach a stage where they can direct their energies to the creation of a scientific tradition that is independent of European science and productive in its output. Stage three science is characterized by the greater number of scientists trained and working at home, their ability to communicate with nationally as well as internationally placed scientists, the prestige attached to their profession, and their ability to sustain scientific effort over time. Basalla suggests that both the United States and Russia reached stage three science between World War I and World War II, equaling if not surpassing European science by this time. The science of Japan, Australia, and Canada he places a rung below that of the United States and Russia. Some of the countries of Latin America and Africa occupy a position yet lower, showing much potential for future growth, yet with major obstacles in the way of establishing independent scientific traditions.

One fault with this model is that it tends to ignore the high degree of scientific interdependence among the advanced scientific nations of the world.[3] Basalla also fails to consider the fact that, for the developing world, the *progression* from stage two to stage three science may well entail an increasing dependence rather than independence on science and technology outside the country, just as industrialization in the developing world has often carried with it increased reliance on foreign materials, technology, and support. There may indeed be a stage between stage two and stage three, a stage of partial independence and partial dependence, which will be most characteristic of many of the countries in the process of development. Countries that come late to industrialization and are already dependent on the science and technology of the industrial world may encounter special impediments during the transition to autonomy in science, despite the continued spread of scientific activity. The development of the theory of dependency has put the problem of the spread and autonomous growth of science and technology in developing countries in a new perspective, as I show in Chapter 8.

The identification of a "colonial" stage of science is, however, most suggestive. Basalla notes that the quality and character of colonial science will vary from country to country, and will depend to a large extent on the quality and character of science in the transmitter country.[4] The quality of colonial science will also depend, I suggest, on the quality and range of scientific institutions available for imitation in countries other than the primary transmitter country. In the case of the United States, for instance, England was one of the leaders of the scientific movement when colonization of the United States began in the seventeenth century. This fact gave efforts in science in North America a vigor unequaled in any of the Latin American colonies. The Royal Society, granted a Royal Charter in 1662, was founded independently of the still medieval universities, and it propagated the new experimental sciences both inside and outside England. Members of the Society requested information about North American flora and fauna from colonists, published the work of colonial scientists in European journals, elected colonials to membership in the Society, sent them instruments and books, and acted as an agent putting individual colonial scientists in touch with each other. One result was the creation of a "natural history circle" that effectively linked North American with European science and established a tradition of scientific work in the New World.[5] As British science declined in the late eighteenth century, and the American revolution separated America from England, French, and at a later date German, science began to provide new models and continuing contacts for "colonial" science in the United States.

Colonial Science in Spanish and Portuguese America

Turning to the Spanish colonies in the New World, we also find considerable scientific activity in the sixteenth and seventeenth centuries. Colonization by Spain began well over a century before that by the British and significantly, well before the start of the scientific revolution of the middle sixteenth and seventeenth centuries. Conquest of the native populations and consolidation of the position of the Spaniards had already been achieved before Copernicus published his *De revolutionibus orbium coelestium* in 1543. The Spanish Crown responded to requests by colonists to found the first university of the New World; the Royal and Pontifical University of Mexico opened in 1551. This was followed by the establishment of the Royal and Pontifical University of Lima in 1572, and another eight universities were begun in Spanish America before a single one was authorized in the British colonies in America. From the Spanish colonial universities an estimated 150,000 students were graduated before the end of the eighteenth century.[6]

Contemporaneously with the erection of a fairly complete academic structure in Spanish America, Spaniards trained in Europe began the scientific study of native flora, fauna, and peoples. The Spanish Americans, unlike the British colonists, were in touch with all three of the great civilizations of the New World, the Maya, the Aztec, and the Inca. Each civilization received its own scholarly study—the Mayas by Diego de Landa, the Aztecs by Bernardino de Sahagún, and the Incas by Bernabé Cobo.[7] Native herbs were incorporated into the European pharmocopeia, such as tobacco, guaiacum and cinchona. The Spanish Crown also collected information on the habits and customs of the Indian people, as well as on the local climate and geography. Questionnaires were sent by the Crown to the local authorities in colonial towns between 1577 and 1586, resulting in the *Relaciones Geográficas*, an important source of data.[8]

This promising beginning in science in Spanish America was checked in the seventeenth century by the Counter-Reformation, which resulted in the imposition of a rigid intellectual orthodoxy in Catholic communities. Science increasingly became a property of Northern Europe and of Protestant countries. With the trial of Galileo in 1633 the Church effectively put itself in opposition to the spirit of free inquiry that lay at the heart of the scientific revolution then gathering momentum. The Counter-Reformation was felt especially strongly in Spain and Portugal. In the Spanish colonies ecclesiastical authority and privilege were extended through the Church's control of land and education. New works in science were often prevented from reaching colonists through the Church's control over the supply of books. Colonial scientists of the stature of the seventeenth century Mexican savant Carlos de Sigüenza y Góngora, whom Irving Leonard calls a "baroque incarnation of the Renaissance man of learning," or the Peruvian historian, linguist, mathematician and astronomer, Don Pedro de Peralta Barnuevo y Rocha y Benavides, were exceptions.[9]

In defense of Latin American academic learning, Lanning demonstrates that the backwardness of Spanish American colonial universities has often been exaggerated. Although Aristotelean physics and metaphysics continued to dominate the arts degree until late in the eighteenth century, the actual lag between new ideas in Europe and their reception in Spanish America was often quite small. By 1736, for example, when the French scientist La Condamine visited Spanish America, the works of Descartes, Leibniz, and Newton were all being taught in Quito. By the middle of the eighteenth century, the schoolmen were under direct attack, and Enlightenment thought began to have an impact. Moreover, Lanning suggests that throughout the colonial period the Spanish American universities provided a thorough clerical training, as good as, if not better than that provided in the colonial universities of the British colonies, an education that played a fundamental role in the maintenance of the hegemony of the Church in intellectual life.[10]

The great difference between the areas of the New World colonized by Spain and Portugal and those colonized by the French and the English lay less in the quality of university education, than in the relative absence in Spain and Portugal of secular academies and societies dedicated to the promotion of the new experimental sciences. It was the academies such as the Royal Society in England that spearheaded the scientific revolution and played a key role in extending science to areas outside Europe. The absence of such societies was crucial for the slow development of interest in the sciences in Latin America. Moreover, the number of institutions of science outside of Spain and Portugal and "available" for imitation and emulation in Latin America was limited. While British colonial scientists were in touch with English and other European scientists, foreign contacts in Spanish and Portuguese America were reduced not only by the cultural isolation of Spain and Portugal but by the deliberate exclusion of foreigners from the colonies. It was not until well into the eighteenth century that the first full-scale foreign scientific expedition, led by the French mathematician, Charles-Marie de la Condamine, was given permission by the Spanish Crown to enter Spanish territory in the New World and make measurements of the meridian of an arc of a degree of latitude, in order to resolve a dispute between Newton and French scientists about the shape of the earth. Accompanied by several scientists, including two from Spain, La Condamine's travels in what are now Peru, Ecuador, and Bolivia lasted ten years, from 1735 to 1745.[11] After him the pace of foreign exploration of Latin America accelerated. The great explorer and naturalist

Alexander von Humboldt was especially important in opening up the wonders of the continent to the imagination and curiosity of European scientists, through his descriptions of the rich tropical floras and faunas along the Orinoco and Negro rivers and in Mexico, Bolivia, Colombia, Ecuador, and Peru. Humboldt was responsible for establishing a long tradition of scientific exploration of Latin America by Europeans, a tradition that has persisted well into the twentieth century.[12]

The re-opening of stage one of exploration science by foreign scientists in the eighteenth century revived the interest of the Spanish Crown in learning more about its overseas possessions. Under the Bourbon King Carlos III, who was on the throne of Spain between 1759 and 1788, patronage of the arts and sciences increased and a number of scientific expeditions were organized. The botanists Hipólito Ruiz López and José Pavón led an expedition to survey the flora of Peru and Chile in 1777, and the specimens brought back to Spain in 1788 were added to the collections of the Royal Botanical Garden of Madrid. Ruiz prepared his *Flora peruana et chilensis*, four volumes of which were published. The 1780s saw the Royal Scientific Expedition to New Spain, as well as the Around-the-World Expedition of Alejandro Malaspina.[13]

While the pace of exploration continued, the history of science within Spanish America until the opening of the nineteenth century belies any steady progression from stage one through stage two to stage three science. Growth was instead interspersed with periods of stagnation. The active period of the sixteenth century, when universities were founded and many works in natural history and anthropology published, was followed by a century or more of decline of activity in science. In the middle of the eighteenth century, partly under the influence of the Enlightenment, especially Linnaeus' work in botany, there was a renewal of interest in science in Spain and the colonies, which led to the founding of new institutions and the reorganization of the old. At some point in the nineteenth century, some fields of "colonial" science reached new levels of output; in other fields political events checked the spurt in science at the end of the eighteenth century, and a "colonial" tradition only emerged slowly and unevenly.

The causes of the persistence of a colonial tradition of science in Spanish America long after political independence are complex.

Basalla rejects cultural imperialism, "whereby science in the non-European nation is suppressed or maintained in a servile state by an imperial power," as a serious factor.[14] He emphasizes instead the small value attached traditionally to science, the small size of the scientific community, and the failure to establish a new role for science as an autonomous intellectual activity, rather than any deliberate effort by foreign powers to suppress efforts to undertake original scientific work. In Spanish America a colonial tradition persisted in part because the independence movement did not result in a sharp break with the past. As Stanley and Barbara Stein observe, the political elite in Latin America sought not to overturn colonial society, but to replace Spanish and Portuguese authority with their own, while preserving the colonial heritage and political and social structures.[15] Colonial patterns of social stratification were in fact consolidated in the first decades of independence. In the United States, by contrast, although psychologically a feeling of dependency upon European culture persisted, independence also created a sense that a "new nation" was in the making that would eventually forge its own culture and values independently of Europe. In Latin America, however, continuity rather than discontinuity with the past characterized the first decades of the nineteenth century, and the Steins argue that economically, at least, it was not until the late nineteenth century that Latin American nations freed themselves in any significant sense from economic dependency on Europe. The leading institution of colonial culture, the Church, maintained many of its traditional structures and privileges in Latin America, as well as its basic role in forming values, until well into the nineteenth century, and this also was a factor in the persistence of attitudes and values characteristic of the past.

Science in Brazil followed to a large extent the pattern already described for Spanish America. The same factors that had been at work in Spanish America to limit science—the poor quality of science in the transmitter country, a policy of cultural imperialism in the seventeenth and eighteenth centuries, and the lack of a decisive break with colonial traditions at the time of independence—helped shape the evolution of colonial science in Brazil. Differences between Brazil and the countries of Spanish America were the result of the greater isolation of Brazil from the rest of the world, and the greater degree of continuity with colonial culture in the nineteenth century, owing to the way in which independence from Portugal came about.

As in Spanish America, the first to study the fauna, flora, and peoples of Brazil were explorers and colonists. One of the first to make scientific observations was Mestre João, who accompanied Cabral to Brazil and in a series of letters to Portugal described the animals and flowers he had encountered. In one of the letters he also reported the first determination of latitude made in the country.[16] In the seventeenth century Portugal was increasingly isolated from the mainstream of scientific thought in Europe. Portuguese efforts in science in the seventeenth century cannot compare, for instance, with those of the Dutch in Brazil. Under the leadership of Prince Maurice of Nassau, in the Dutch colony of Pernambuco in the northeast, a number of scientists were brought to Brazil between 1630 and 1654. The first observatory was established in the New World under the Prince's patronage.[17] The naturalist George Marcgraff and the physician Wilhelm Piso began a systematic study of Brazilian fauna and flora. Their work resulted in the publication, in 1648, of the great *Historia naturalis brasiliae.* The Dutch experiment with science in the New World ended in 1654, when the Dutch were expelled from Brazil by the Portuguese.

While Dutch experimental science thrived in Brazil in the seventeenth century, Brazilian efforts were limited by the mercantile policies of the Portuguese Crown, which viewed Brazil primarily as a source of raw materials for export. By the end of the seventeenth century, trade with the East Indies had declined, and Portugal increasingly relied on the tobacco, brazil-wood and, above all, sugar, that Portuguese and British merchant ships brought from Brazil to Lisbon for distribution all over Europe. Gold was discovered at the end of the seventeenth century in Minas Gerais, providing a new source of wealth for Brazil and Portugal until the middle of the eighteenth century. The discovery of gold by roving bands (bandeiras) brought about a shift of Brazil's population away from the coast toward the interior where gold was to be found. It led, notes Boxer, to a neglect of agriculture in favor of mining, and increased the general prosperity of the colony, until by 1750 Brazil was in many ways more prosperous than Portugal.[18]

In keeping with Portugal's interests in raw materials, most efforts in science in the eighteenth century were directed to collecting information about new products of possible commercial value. The governors of the Brazilian captaincies collected specimens of plants and animals. Boundary disputes and the need for accurate maps stimulated a tradition of topographical and cartographical work. Military engineers proved competent in surveying and map-making. The major rivers were explored sporadically throughout the eighteenth century.[19] Unlike the Spanish, however, the Portuguese did not transfer to the New World the educational and cultural institutions of the Old. No universities as such were founded until the twentieth century, and formal medical education in Brazil was delayed until 1808. Almost all education above the secondary school level took place in Portugal. Within Brazil, the Jesuit missionaries remained the chief transmitters of European knowledge until 1759, but their efforts to expand their educational activities were prevented by the Portuguese authorities, who feared the establishment of any institutions in Brazil that might rival those of Portugal. The request of the Jesuits and the public to elevate the status of the Jesuit College in the old capital of Bahia into a university was denied, on the advice of the Portuguese University of Coimbra.[20] Printing presses, which might have brought new ideas into Brazil, were systematically suppressed.[21] Institutionally, therefore, science in Brazil was considerably behind that of Spanish America until the end of the colonial period.

The year 1750 marked somewhat of a turning point in Brazilian history. It signaled the beginning of the end of Brazil's golden age, as the supply of gold began to decrease. It also marked the rise to power of the remarkable Marquis of Pombal, whom Boxer calls "the virtual dictator of Portugal."[22] In 1759 Pombal expelled the Jesuits from Brazil, and many of the schools in Brazil closed. Countering this blow to the educational structure within Brazil, Pombal, partly under the influence of the Enlightenment, overhauled the medieval curriculum of the University of Coimbra, and established several new chairs of science. At government expense, the Brazilian mineralogist José Bonifácio de Andrada e Silva was sent, together with two companions, on a scientific journey to Europe to improve his knowledge of geology, mineralogy and particularly the art of assaying. In 1783, the government also sent the Brazilian-born naturalist Alexandre Rodrigues Ferreira, who had studied at the recently reformed University of Coimbra, to explore the fauna and flora of Brazil. In a long journey lasting several years, Rodrigues Ferreira collected many specimens and made valuable observations, particularly of the indigenous Indian tribes. Unfortunately his collections were scattered after his return to Portugal and a large number of his manuscripts remained unpublished. The work of the Brazilian-born botanist, Frei José Mariano da Conceição Veloso (1742-1811), who spent many years studying the flora in Brazil, should also be mentioned; his *Flora fluminensis* appeared in eleven folio volumes between 1825 and 1827, and his *Quinographia portugueza* in 1799, both in Portugal.

Enlightenment emphasis on useful learning also began to have its influence in the formation of new scientific societies both in Portugal and in Brazil. The *Scientific Society of Rio de Janeiro* (Sociedade Scientífica do Rio de Janeiro), founded in 1772, was in marked contrast to the earlier, usually ephemeral, literary and historical societies in Brazil. The Scientific Society was designed to spread scientific knowledge, and its first meetings were attended by four surgeons, three physicians, two apothecaries, and a farmer. The fields of botany, zoology, chemistry, physics, and mineralogy were all represented, and public lectures offered. A small botanical garden was started, where experiments on plants were carried out. In 1779 the society changed its name to the *Literary Society of Rio de Janeiro* (Sociedade Litterária do Rio de Janeiro) but continued its scientific work until 1794 when the society was closed, probably for political reasons.[23] Botanical gardens were also founded in the provinces of Brazil in the same spirit of Enlightenment enthusiasm for practical knowledge.

Science in Nineteenth Century Brazil

The groundswell of interest in science occurring in the last years of the eighteenth century received a great boost in 1808, when Brazil underwent a profound transformation owing to the transfer of the Royal Court from Lisbon to Rio de Janeiro. Late in 1807, the Portuguese government found itself threatened by the advance of Napoleon's troops toward Lisbon. When the Portuguese Prince Regent, Dom João, refused to aid Napoleon's efforts to blockade British ships in Portuguese waters, the Court was forced to flee the country for its American colony. Accompanied by his aging and mad mother, Maria I, and by a crowd of court administrators, noblemen and their families, military officers, and clergymen, Dom João set off for Brazil in an assortment of military and merchant ships, arriving in Bahia, the old capital of Brazil, in November of that year. After a stay of a few weeks, the court then made its way south to Rio de Janeiro, and established its residency there in March 1808. In 1815, Brazil was elevated to the status of a kingdom, and the colonial country became the administrative headquarters of the Portuguese Empire in a transfer of power from mother country to colony virtually unprecedented in history.[24]

Brazilian life was dramatically altered by this political event. One of the first acts of the Prince Regent was to open the ports of Brazil to European trade, on the advice of the governor of Bahia, and of the economist José da Silva Lisboa, a follower of Adam Smith. The city of Rio de Janeiro, with a population of about 80,000, came to life as ships from Europe called in at the port bringing the latest manufactured goods, European books, and Portuguese and European visitors. A printing press was established for the first time since the abortive presses of the eighteenth century. A royal library of 60,000 volumes was opened to the public. Steps were taken to redress the situation created by the absence of educational facilities in Brazil and the difficulties of continuing to rely on a supply of doctors and engineers from Europe. By 1838, the Historical and Geographical Institute of Brazil (Instituto Histórico e Geográfico Brasileiro) had been founded, and in later periods science was stimulated by various agricultural, medical, and other societies. Virtually overnight, opportunities for the advancement of science in Brazil were increased.

Of most pressing concern was the shortage of medical doctors. The Prince Regent's personal physician, Dr. José Correia Picanço, a Brazilian-born professor from the University of Coimbra, urged the authorization in 1808 of medical and surgical courses in the military hospitals of Bahia and Rio. Later the courses were regularized as Academies of Medicine and Surgery, and the professors given the right to teach medicine and organize clinics. New legislation in 1832 led to the formal creation of two Medical Schools, each with fourteen professors and six substitute professors, and with an academic curriculum closely modeled on that of the Faculty of Medicine in Paris. Examination and licensing of doctors were left in the hands of the Medical Schools and the local municipalities, with the Schools setting the examinations for the title of doctor, and the municipalities certifying the titles and issuing the licenses for the practice of medicine in the district. A lively medical press was soon established. Licensed and unlicensed druggists, as well as faith-healers of all kinds, served the medical needs of those unable to afford or out of reach of licensed physicians.[25]

Meanwhile, the beginnings of regularization of public health were made by the organization of the traditional Portuguese institution of Chief Physician (*físico-mor*), and Chief Surgeon (*cirurgião-mor*) who were responsible for nominating delegates to the provinces to oversee public health and regulating the sale of alcoholic beverages, as well as issuing regulations concerning the sanitation of the city. The institution of *físico-mor* was abolished in 1827 because of its ineffectiveness, and a committee on public hygiene established in its place to inspect foreign ships in port. The inadequacy of sanitary regulations led the Society of Medicine in Rio de Janeiro to undertake a study of hygiene, and its recommendations were put into law in 1830. For many years much of the law remained a dead letter. However, the Medical School together with the Medical Society of Rio, continued to act as useful sources of advice and support in times of epidemics and other crises in public health.[26]

A second area where the need for trained people was felt was military engineering. In 1810, a Military Academy was founded to train officer cadets in the arts and sciences of war, and prepare them for the surveying and exploration of what was still a virtually unknown land. The founding of this institution, it has been suggested, represented a deliberate effort by the Prince Regent to alter the traditional literary mentality of the country.[27] The most modern European textbooks in mathematics and physics were imported for use in instruction—works by Euler, Bezout, Monge, Legendre, Lacroix, and Haüy. Later these ceased to be read in the original but were studied in compendia put together by instructors at the school.[28]

Efforts to provide an institutional foundation for natural history came a little later. The case of natural history in Brazil in the nineteenth century raises a number of interesting questions about which sciences get established first, and why, in a developing country. Natural history flourished in the nineteenth century in the United States because exploration and mastery of the wilderness were an essential part of the North American experience, and because the continent was rich in data of interest to European scientists. American naturalists therefore had a comparative advantage in this field.[29] There are other reasons why natural history might have been expected to develop in Brazil. Like the United States, Brazil was also vast and still unexplored, as well as rich in minerals, plants, and animals of every conceivable variety. The study of natural history at that time did not depend, as did physics, on sophisticated mathematical training, and therefore did not depend on the evolution of institutions of secondary education. It was a field to which amateur naturalists could make contributions of an empirical nature, before more sophisticated systems of natural classification encouraged specialization and increased professionalization.[30] Moreover, though the usefulness of natural history was not as patently obvious as that of medicine and engineering, the charting of mineral deposits, the discovery and collection of plants and animals that might be consumed or exported, and the cultivation of foreign plants might all have been expected to appeal to the utilitarian interests of the national government. Daniels, in his study of science in Jacksonian America, notes that regardless of the actual benefits derived from natural history, belief in the utility of its study was widespread in the nineteenth century.[31] Furthermore, in Brazil natural historians were less isolated from European scientists than were the Brazilian students of the exact sciences, owing to their contacts with foreign naturalists present in the country throughout the century. Indeed, developments in natural history were strongly influenced by foreign scientists, though they did not match the expectations suggested by a consideration of the factors analyzed here.

The tradition of scientific exploration of South America established by Humboldt gained momentum in Brazil with the opening of Brazil to European trade after 1808. Many expeditions were sponsored, some privately and others by foreign governments. The travels of the French naturalist Auguste de Saint-Hilaire in 1816 were followed by those of Alcide d'Orbigny, who was sent by the Muséum d'Histoire Naturelle de Paris, and by those of the German Prince Maximilian of Wied-Neuwied, who was accompanied by the botanist Friedrich Sellow. With the marriage of the Archduchess Leopoldina, daughter of the Austrian emperor, to the Brazilian Prince Regent, Dom Pedro, a number of scientists came to Brazil with her court to examine Brazilian vegetation and animals. Most famous were two Bavarians, Karl Friedrich Philipp von Martius and Johann Baptist Von Spix, whose massive, many volumed *Flora brasiliensis* (the first volume of which was published in 1829) eventually took sixty-six years to complete, and remained the standard textbook on Brazilian botany until well into the twentieth century. In the steps of the French and Germans came the Russian-sponsored expedition of Baron Georg Heinrich von Langsdorff, a German diplomat in the service of the Tsar, who collected a herbarium of 60,000 specimens for St. Petersburg. The

English were well represented with the visits to Brazil of Charles Darwin, Henry Bates, Alfred Russel Wallace, and the botanist Richard Spruce. American science began its own tradition of scientific exploration in Brazil when the Thayer expedition, led by the distinguished Swiss-born zoologist Louis Agassiz, came to Brazil in the winter of 1865–1866 to explore the Amazon. This stage of exploration science led to the amassing of a large amount of important scientific data.[32]

The first national institution in natural history to be created was the Royal Garden, later called the Royal Botanical Garden. Here it was hoped to acclimatize foreign plants for their use in Brazil and for their beauty. Tea plants were grown and tea sold in Rio, although the experiment did not last. The cultivation and study of native Brazilian plants was neglected for a long time.[33]

The Imperial Museum of 1818 was founded partly in response to the interest in natural history aroused by the work of foreign naturalists and partly in cultural imitation of the great museums of natural history of Europe. Its purpose was to "spread knowledge and the study of natural history in the kingdom."[34] The botanist Sellow was charged by the Brazilian government with collecting specimens for the museum during his travels. The provinces cooperated with the museum by sending materials on request. By 1842, the collections had grown to such an extent that the museum had to be organized into departments, and its rooms and cabinets increased. Nevertheless, by 1844 the budget was cut back to a quarter of its former size, and the museum's director commented sadly that "the usefulness of our museum is not yet perfectly understood by our national Congress, nor have a large part of our administrators realized the beneficial influence of similar establishments."[35] Despite this setback, interest among Brazilians was sufficiently strong to lead to the founding of a natural history society in 1850, the *Sociedade Velosiana de Ciências Naturais*, named after the eighteenth century botanist Father José Mariano de Conceição Veloso. Attending the society's sessions were a number of scientists from the Imperial Museum, as well as the director of the Botanical Garden. Meetings continued until 1855, when the society came to an end, only to be replaced the following year by another, the *Palestra Científica*. This proved equally short-lived, but nonetheless the activities of the naturalists in Rio de Janeiro stimulated the prestigious Historical and Geographical Institute of Brazil (Instituto Histórico e Geográphico Brasileiro) to suggest to the government the formation of a scientific commission to explore Brazil. The government accepted the proposal and a commission was appointed which lasted two years. Some good scientific work was carried out, but disputes among the commission members led to its termination and much of the scientific work was left unpublished.[36]

The fact was that, although European interest in the natural history of Brazil was increasing, Europeans working in Brazil could not, by themselves, overcome the traditional Brazilian indifference to science. The careers of most of the foreigners took place independently of Brazilian institutions, and their communication was primarily with centers of science outside Brazil. In the few instances where foreign scientists spent a large part of their working lives in Brazil, as was the case with the brilliant German biologist Fritz Müller, and the gifted Danish paleontologist Peter Wilhelm Lund, their small effect on the natural history tradition within Brazil reflected the general lack of support for non-utilitarian science, the scientists' geographic distance from the major cities, and the weakness of existing Brazilian institutions.[37]

Generally speaking, there was little in Brazilian society to encourage the disinterested pursuit of science or the development of applied science and technology before the last decades of the nineteenth century. Although the country experienced a number of formal political changes in the first decades of the century, including separation from Portugal followed by independence in 1822, the monarchy was preserved until 1889 and stood as a symbol of continuity with Brazil's colonial past. The monarchy helped preserve the power of the traditional social and economic elite and their values. While the cities of Rio de Janeiro, Bahia (Salvador), and São Paulo grew rapidly, especially in the last third of the century, the heart of Brazilian society was still the large plantations and the export economy they supported, based on sugar, and, after 1840, coffee.[38] Brazilian culture was predominantly agricultural, hierarchical, and patriarchal. The persistence of slavery until 1888 may also have contributed to the survival of a plantation society that made few demands for science. The transport revolution got underway in Brazil slowly and only in the twentieth century did railroads begin to link together effectively the different regions of the vast country.[39] Immigration, which accelerated rapidly after 1888, when plantation owners sought to find a substitute for slave labor, did not initially disturb the rhythm of Brazilian life.

Many immigrants were drawn from southern Italy and were without technical skills or education. They were quickly absorbed into existing social structures. Industry, too, developed slowly, impeded in part by the absence of limited liability laws before 1888. Secondary education was a privilege enjoyed by the few, not a right for the many. The emphasis in education was literary rather than scientific, reflecting the interests of the elites. If science did not develop in nineteenth century Brazil, in short, it was because little value was attached to the pursuit of science for its own sake, and industrialization had not progressed to the point where it could provide new sources of support for utilitarian and practical science.

Science in the Old Republic

In the last three decades of the century, changes occurred in the political and social life of the nation that had an effect on science, and prepared the way for further growth in the twentieth century. A new generation of intellectuals came to the fore in the 1870s to question the efficiency of monarchical government, the morality of slavery, and the quality of Brazilian culture. Partly through their efforts, slavery was finally abolished in 1888, Brazil being the last country in the western world to take this step. The end of slavery was followed by the collapse of the monarchy in 1889, the exile of the Emperor, Dom Pedro II, and the formation of the Brazilian Republic. In the field of education, the brilliant polemicist Rui Barbosa published a plan for the reform of the entire structure of secondary education. The need for a university, long a theme among intellectuals and educators in Brazil, was raised again in discussion, although no university was to be founded until well into the twentieth century. From the Rio Medical School the bacteriologist Dr. Domingos Freire was sent to Europe in 1877 to study the best ideas in medical education.[40]

Changes in science occurred piecemeal and were stimulated not only by the events described above but also by a variety of other factors. Brazilian doctors and scientists, many of them trained in Europe, were among the most important spokesmen for an improvement in Brazilian science. Foreign scientists played a part in bringing to Brazil European ideas and organizations. The threat of epidemic disease began to stimulate local and federal authorities to organize public health agencies, as medical science advanced. Economic and industrial development, such as in the areas of electric power and railroads, began to create new sources of employment for engineers in the last years of the century.

In an interesting article written in 1883 for the new American journal *Science*, the American geologist, Orville Derby, who was working at the National Museum in Rio de Janeiro, summed up the period of the last ten to fifteen years as one of an "awakening to the importance of scientific research." He attributed this awakening to the widening of contacts between Brazil and other countries, increased communications within Brazil, to the new energy in national life brought about by the war with Paraguay in 1870, the visit of the American naturalist Louis Agassiz with the Thayer expedition in 1865–1866, and in particular to the Emperor Dom Pedro II's first trip to Europe and the United States. While in the United States, the Emperor, himself an amateur scientist, visited many scientific museums and schools. His trip abroad, following on Agassiz' visit, led to a new determination on the part of the Brazilian authorities to sponsor scientific development.[41]

Of prime importance was Louis Agassiz' visit to Rio de Janeiro. Agassiz impressed Brazilians as the epitome of the cultivated man of science, just as he had impressed Americans when he arrived in Boston from Europe in 1846. He gave a series of well-publicized lectures at the Museum, attended by the Emperor, and, on the insistence of Agassiz and his wife Elizabeth, by a number of women. He visited several of the leading scientific establishments in Rio de Janeiro, and the comments in his book, *A Journey in Brazil*, published in 1868, provided Brazilians with some measure of how their scientific establishments compared with those of North America and Europe. On the whole, Agassiz found the Brazilian scientists lacking in an interest in experimental science and their institutions inadequately supplied with materials for undertaking modern science. "Surrounded as they are with a nature rich beyond compare," he wrote, nonetheless, "their naturalists are theoretical rather than practical." He attributed the absence of experimental science in part to the institution of slavery, which he believed led to disdain for the manual labor that was essential for science. "As long," he wrote

> as students of nature think it unbecoming a gentleman to handle his own specimens, to carry his own geological hammer, to make

> his own preparations, he will remain a dilettante in investigation. He may be very familiar with recorded facts, but he will make no original researches.[42]

Of the various institutions he visited, he judged the Military Academy approximated most closely a proper scientific institute. Even here, however, he was struck by

> the scantiness of means for practical illustration and experiment; its professors do not yet seem to understand that it is impossible to teach any of the physical sciences wholly or mainly from books.[43]

The Imperial Museum he described as "antiquated." Of its specimens of fish he remarked that "a better collection might be made any morning in the fish market."[44]

Just as important to Brazilian science as Agassiz' own visit was the effect of the young American scientists he introduced to Brazil. The geologist Frederick Hartt, an original member of the Thayer expedition, returned to Brazil in 1867 to complete his geological studies, which resulted in *The Geology and Physical Geography of Brazil* in 1870, the first general textbook of Brazilian geology. Hartt's student at Cornell, Orville Derby, was brought to Brazil by Hartt that year, together with a number of other students, and in 1875 the Brazilian government accepted a proposal made by Hartt the preceding year for an imperially sponsored geological survey, of the kind familiar in the United States. The Imperial Geological Commission, of which Derby was a member and which Hartt directed, undertook important topographical studies in the areas surrounding Rio de Janeiro. Unfortunately, the government terminated the Commission in 1877, while the Emperor was out of the country. The following year Frederick Hartt died in Rio de Janeiro at the age of 38, a victim of yellow fever.[45]

Meanwhile, the desire of young Brazilian scientists for careers in science within Brazil was on the rise. This was especially true of those returning following European training in science. In the field of natural history, the young Brazilian botanist Ladisláu Netto, for instance, came back to Rio de Janeiro in 1866, the year of Agassiz' visit, only to find himself

> discouraged as would anyone be who felt a desire to maintain

> himself in contact with the scientific movement in Europe and who found himself in a position which made it almost impossible to take a single step to satisfy this normal ambition.[46]

Ladisláu Netto was appointed chief of the botany section in the Museum, and publicized the deplorable conditions in the museum in a sarcastic report in 1870.[47] As a result he was made vice-director, subsequently director, and was able to greatly expand the facilities of the museum. New departments were established, and several foreign scientists appointed to positions in the museum. Among these were Frederick Hartt, at the time working with the Imperial Geological Commission, and after his death, his colleague, Orville Derby, who headed the geology section. The Swiss naturalist Émile Goeldi, and the French industrial biologist, Louis Couty, were also among those joining the staff. Other foreign naturalists already working in Brazil were attached to the museum through the position of "travelling naturalist." In this way, the scientific work of Fritz Müller, and of the zoologist Hermann von Ihering, who had resigned a chair at the University of Jena to study the fauna of Brazil, was associated with the museum.[48] The young Brazilian physician João Batista Lacerda, who had worked briefly at the Misericórdia Hospital in Rio and was restlessly looking for a place to carry out research, was brought to the museum to organize its anthropological work. An anthropological exhibition was held in 1882, and several papers published in the newly founded journal of the museum, the *Archivos do Museu Nacional*, the first issue of which had appeared in 1876.[49] For many years this journal remained the only significant journal of scientific work (as opposed to medical research) in the country. The museum also embarked on a new direction when Lacerda and Louis Couty joined forces to start a physiological laboratory, where they undertook experiments on the physiological effects of curare and other poisons.[50]

Similar reforms took place in the Military and the Medical School in Rio de Janeiro. The separation of military and civilian engineering had begun in 1858, when the first degrees in civil engineering were offered, and was completed by the creation of two schools. The Central School continued to train military officers, and a Polytechnic School was founded in 1876 as the first full-fledged civilian engineering school in the country. A Mining School was also started in Ouro Prêto that year. A Polytechnic School was also begun in São Paulo in 1893.

These developments reflected the growing incentives for industrial engineering. In the Rio Medical School, a blistering attack on the school's facilities made in 1879 by the director, Dr. Nuno de Andrade, finally stimulated the government to action.[51] Under the direction of the Viscount of Saboia, several reforms were finally undertaken, including the building of laboratories, which Saboia boasted in 1883 rivaled the best in Europe.[52]

Needless to say not all these steps in the direction of improving science were sustained. At the National Museum, tensions between the director, Ladisláu Netto, and his staff were often acute. Many foreigners left Brazil, dissatisfied with the opportunities for research. Another loss occurred when Émile Goeldi and Hermann von Ihering left to become directors of the new museums in Pará and São Paulo, respectively. Paradoxically, the growing interest in science and the demand for scientific administrators drew some of the most talented scientists away from research and hurt the scientific community emerging in Rio de Janeiro by scattering its members to areas geographically and intellectually isolated from each other. This scattering is in fact a recurring problem—the creation of a diversified community destroys the chance to create a "critical mass" of scientists. The place of the physiological laboratory within the museum was another vexing question, and Ladisláu Netto's disagreement with Lacerda over this matter was one factor in Ladisláu Netto's eventual resignation. Lacerda's sincere but misplaced search for the "bacillus" of yellow fever also probably hurt the reputation of the museum.

The Polytechnic School, though initiating an important tradition in engineering, was nonetheless confined to the role of high-level pedagogy. The lack of books in physics and mathematics, the traditional dependence on theory rather than empirical observation and experiment, and the small demand for civilian engineers all resulted in the fact that the community of mathematicians and physicians was too small to carry out any original work. In the United States, where engineering schools also developed slowly, nonetheless engineers trained at West Point supplied the manpower for new industries being founded, and eventually a number of civilian engineering schools, such as the Massachusetts Institue of Technology, were established. This incentive for broad-based engineering science was absent in Brazil. The history of the Polytechnic School in

Brazil suggests that a clear distinction must be made between an institution involved in transmitting scientific ideas taken from elsewhere and an institution forced by the demands placed on it by society to make new discoveries and play an active role in the international republic of science.

The spurt of interest in geology shown in the establishment of the Geological Commission in 1875 and the founding of the Mining School in 1876 was not followed by a consolidation of the institutional base of geology, although American geologists continued to show considerable interest in Brazil.[53] The history of the geological sciences in this period is, in fact, extremely valuable for raising several questions about the role of foreigners in stimulating scientific advance in a developing country, and the supports for science of potentially practical use in a country with a small industrial base. When Frederick Hartt came to Brazil in the late 1860s, the time seemed ripe for the advance of geology. The national government had responded to Hartt's proposal by sponsoring the first proper national geological survey. Furthermore, the state of geology was such that discoveries made during surveys and explorations carried out for practical and economic purposes were often seen, either at the time or retrospectively, to have important implications for geological theory. In effect, given a leader sensitive to the political need for practical results, support for economic geology could become support for theoretical geology. Here the experience of the geologists in the U.S. Geological Survey is instructive. Under John Wesley Powell in the 1880s, geologists carrying out surveys for the federal government were able to make important contributions to the theory of geological change, to Pleistocene geology, and to invertebrate and vertebrate paleontology.[54]

In Brazil, however, a variety of economic, professional, and personal factors combined to prevent this from occurring. Hartt's role in Brazilian geology was taken over by his former student and then colleague in the Geological Commission, Orville A. Derby, following Hartt's untimely death in 1878. Derby stayed in Brazil for more than forty years, first in the National Museum, and later, after the state governments took up the task of sponsoring geological survey work, in the state survey of São Paulo. His work, when coupled with that of the French mineralogist Henri Gorceix at the Mining School at Ouro Prêto, made Brazil a leader in geology among the countries of Latin America.[55] Yet in the long run, industrial development in this period was too slight to spur any rapid development of the Mining School. New industries such as railroads still tended to rely on foreign technicians and equipment.[56]

In the area of surveys, the situation called for a judicious balance between practical and theoretical geology that in practice was difficult to work out. As director of the state geological survey of São Paulo, Derby showed a distaste for administration and often failed to satisfy the legislature's quite natural concern for some practical return for their investment in geology. Reports were delayed while Derby collected more and more data; indeed, no comprehensive account was presented to the legislature for ten years. While the state was primarily interested in economic and structural geology, Derby's chief interest lay in paleontology, which had little immediate importance for the economic development of the state. Derby's inability to satisfy the state's demand for results and his own interest in theoretical geology, as well as the absence of any institutional organization within which the work of the geologists could be consolidated, all probably contributed to the judgment by the Brazilian geologist Victor Leinz that, while all recognized the magnificent work carried out by Derby and his colleagues, "we judge he did little in the way of creating schools of researchers, transferring to them his methods of work."[57]

The history of geology in Brazil in this period also raises the question of the role of the foreign scientist in promoting science in a developing country where science enjoys little prestige. Most of the foreign scientists who came to Brazil in this period to found the institutions with which they were familiar elsewhere (as did Hartt), were, with some exceptions, relatively young and therefore unknown as they embarked on their careers. Once in Brazil, regardless of the fact that, like Derby, many published extremely good work in their fields, the scientists were at a disadvantage in initiating new scientific developments owing to their lack of political influence. In this respect it is interesting to compare Louis Agassiz' career in the United States with that of Derby in Brazil. Agassiz' success in promoting scientific institutions that would provide him with a satisfying career was a function of his great energy as an entrepreneur in science, the greater supports that existed for science in the United States at the time of his arrival compared to those in Brazil, and his marriage to Elizabeth Carey, a member of a prominent Boston family.[58] In contrast, Orville Derby came to Brazil as a young and relatively unknown geologist at a time when the political elite showed almost no interest in the advance of science. Derby was also a bachelor, and lacked the entrepreneurial skills so much a part of Agassiz' role in North American science.[59]

Derby's experience suggests that the work carried out by a foreign scientist in a developing country where few economic or political incentives for science exist may well be less effective than the work of a strategically located national scientist. The situation in Brazil around the turn of the century, that is, may have called for a "Brazilian" solution precisely because the country was less prepared culturally and economically to support science, than was the United States in the 1840s, when Agassiz first arrived there.[60]

The Setting in 1900

To sum up, by 1900 the number of institutions of science, though still small, was on the increase, and the facilities of several of them had been improved. The number of foreign scientists working in Brazil, many of them on contract to the government, had also increased, and they spurred a more disinterested pursuit of science. With the founding of the Republic in 1889, and the passing of responsibility for education and science to the individual state legislatures, concern about sanitation and medicine was growing, particularly in the economically thriving state of São Paulo. Men such as the São Paulo physician and microbiologist Adolfo Lutz, and the sanitarian Emílio Ribas, were rising to prominence. These scientists were fully in touch with professional standards in their fields, and were publishing work in European and Brazilian journals. Opportunities for careers in science were slowly increasing, especially in medicine and engineering.

Nonetheless, despite this "awakening," Brazilian science in 1900 still conformed to the colonial pattern described by Basalla. The scientific establishment was small, and no part of Brazil's educational or scientific structure was capable of producing or training research scientists in a systematic fashion. Originality in science was still a result of individual effort, European training, and, often, private wealth. Lack of funds to travel to Europe to improve training and to increase contacts with those at the forefront of scientific investigation, and the barriers created by Portuguese (sometimes called the "tomb" of thought because so many works written in it have been inaccessible to scholars) were other impediments. The institutionalization of scientific values was far from complete, especially in the government bureaucracies administering scientific institutions. Many problems would have to be dealt with before the colonial tradition could begin to be replaced by a more independent and productive phase of science.

As Basalla notes, in fact, the transformation of his stage two, colonial science, into stage three, independent science is one of the least understood or studied processes in the transference of science to the rest of the world. The analysis of the evolution of science in nineteenth century Brazil has given some indication of the range of problems inhibiting scientific advance. These included the difficulties of giving legitimacy to scientific activity when industrialization was still in its infancy, the problems of administration, the danger of premature fragmentation of nuclei of scientists (such as occurred at the National Museum), and the difficulties of creating large enough communities to carry out satisfactory work.

Basalla indicates some of the tasks he considers must be undertaken before a transformation of colonial science can come about. His list includes the overcoming of resistance to science on religious or philosophical grounds, the clearer definition of the social role of the scientist, the improvement of the teaching of science, the working out of the proper relation between science and the government, the founding of societies to promote scientific effort and open up national and international channels of communication, and the creation of a proper technological base for science.[61] Missing from Basalla's analysis is a discussion of the special effect that the timing and type of industrialization have on the ability of countries to transform science into an independent stage. The fact that many developing countries, such as Brazil in the last decades of the nineteenth century and the first decades of the twentieth century, were only just beginning to industrialize, while industrialization in Europe and the United States was entering a new phase of organization and productivity, led to dependency upon foreign technology, and, in turn, upon foreign science. This state of dependency in itself then became an important factor making the transformation to autonomy in science especially difficult.[62]

Given the situation in Brazil by 1900, it was clear that some tasks would take precedence over others, and that the solution to some of the problems indicated by these tasks would be the result, rather than the cause, of improvements in the position and quality of science. The overcoming of general resistance to science and the improvement of scientific education at the secondary level were closely connected and would have to await profound changes in education, not merely in its content but in its extent. In 1900 fully eighty percent of the population was still illiterate. Among the elite for whom education above the primary level was available, many wealthy plantation owners did not choose to send their sons to the professional schools of law and medicine, which offered the only higher degrees in Brazil. The idea of a general liberal arts education had to await the development of a university in 1920. The system of education characteristic of Brazil in 1900 prepared men for careers as diplomats, politicians, journalists and administrators, but not scientists. The elite in Brazil had little interest in, and little understanding of, modern science. Any institutional development in science therefore, would have to take into consideration the fact that science was still operating in a vacuum and lacked support from education.

The relation between science and the government was certainly another critical area. Unlike in the United States, where a crucial phase of development in science had come about through private patronage, private support for science was lacking in Brazil not because of ingrained anti-intellectualism, but because science and technology were not important to the cultural elite in 1900. After the Civil War, the federal government in the United States began to finance government agencies, such as the Geological Survey, that represented the very best in American science. By 1900, in contrast, Brazilian governmental support of science was still small. The development of a working relationship between scientists and the government in the future and the definition of the proper balance between service-oriented applied science and theoretical science, would be an important issue in a country with limited resources. So would the question of administrative autonomy. It was not clear that by 1900 the federal government was in a position to initiate any fundamental revisions of the state of science in Brazil. On the other hand, any institution of science would have to look to the federal and state governments as the only available sources of support.

It seemed certain that institutions would be at the center of any changes in science after 1900. Yet the past history of the major institutions of science in Brazil was not encouraging. The total number of institutions was still comparatively small, and their role was primarily that of transmitting knowledge developed outside Brazil rather than encouraging original investigation. The lack of professional positions in science often produced rivalries—such as that described between Ladisláu Netto and João Batista Lacerda in the National Museum—that threatened the stability of institutions. The isolation of scientists within Brazilian society, the lack of technical and written materials for scientific work, the absence of adequate financing and of an educational system capable of producing science students suggested that any institution involved in improving the quality of science would have to undertake a large number of tasks simultaneously. These would include the promotion of science, as well as the education of students. The role of the scientific administrator in a society lacking the normal supports for science was therefore of decisive importance. Yet at the same time few administrators or scientists had the necessary institutional experience to guide institutional change of this order.

There were other compelling reasons why the development of science beyond the colonial stage in Brazil would occur at the institutional level. The rise of research science in research institutes, described in detail by Joseph Ben-David, completely changed the needs of science. The types of values, institutions, and bureaucratic arrangements that had produced successful science in 1800 could no longer produce successful science by 1900. Whereas the typical scientist in 1800 invariably worked alone, supported by private patronage or individual wealth, by 1900 success in science depended on the strength of research-oriented institutes supported by government or industry. The typical scientist worked in a laboratory, transmitting his knowledge to students through demonstration and seminars, while carrying out original work of his own. The rise of this type of science, argues Ben-David, was due not only to the acceptance of a research ideal among academic scientists, but also to the rise in industrialization, which created new demands for scientists and technologists. As a consequence, the government and industry became large-scale patrons of science at the end of the nineteenth century in Europe and the United States.[63]

The development of research science in the latter part of the nineteenth century meant that even a partial change in Brazil's scientific output would require not merely greater financial support for science, but also the creation of a completely new tradition of research science. Research science depends not on individual effort, but on the cooperative efforts of many scientists working together and sharing the same ideals of research. It demands technically trained students, long-term financing, and a high degree of academic and bureaucratic autonomy. All these factors suggest the wide range of institutional change that faced Brazil if it were to begin to "count" inernationally in even a single field of science.

Ben-David notes that with the rise of research institutes as the center of scientific productivity, success in science no longer depended primarily on the popular support given science on general grounds. It depended instead on the specific support given, for whatever reasons, to a relatively independent and socially insulated system of research.[64] The history of the sciences in Brazil up until 1900 suggested, in fact, that any future support for a research institute would come from the government, and would be given for utilitarian or nationalistic reasons. This is interesting because it is often argued that a high level of scientific performance must await the development of a very broadly based educational system. The focus on institutions rather than on values, however, meant that scientific advance could come about in the future in Brazil through a process of shrewd institution-building, rather than through a nationwide transformation of values.

References

[1]See, for instance, Arthur P. Whitaker, *Latin America and the Enlightenment* (New York: Cornell University Press, 2nd edition, Great Seal Books, 1961). Other sources of information are the various works dealing with exploration. See references 11 and 12. Recently, Marcel Roche explored the problem of the absence of science in Spain and Latin America in an article which stresses religious factors. See Marcel Roche, "Science in Spanish and Spanish-American Civilization," in Ciba Foundation Symposium 1 (new series), *Civilization and Science: In Conflict or Collaboration?* (Amsterdam: Elsevier, 1973), pp. 143–160.

[2]George Basalla, "The Spread of Western Science," *Science* 156 (May 1967), 611–622. An interesting interpretation of science in Latin America is also given by Irving A. Leonard, "Science, Technology, and Hispanic America," *The Michigan Quarterly Review* 2 (1963), 237–245.

[3]See, for instance, Richard R. Nelson, "'World Leadership,' 'Technology Gap' and National Policy," *Minerva* 9 (July 1971), 386–399.

[4]George Basalla, *op. cit.*, p. 614.

[5]Brooke Hindle, *The Pursuit of Science in Revolutionary America, 1735-1789* (Chapel Hill: University of North Carolina Press, 1956), Chapters 2–5.

[6]John Tate Lanning, *Academic Culture in the Spanish Colonies* (London: Oxford University Press, 1940), Chapters 2 and 3.

[7]Diego de Landa, *Landa's Relación de las cosas de Yucatán*, a translation, edited with notes by Alfred M. Tozzer (Cambridge, Massachusetts: The Museum, 1941); Bernabé Cobo, *Historia del nuevo mundo*. Pub. Por primera vez con notas y otras ilustraciones de Marcos Jiménez de la Espada, 4 vols. (Sevilla: Imp. de E. Rasco, 1890–93); Bernardino de Sahagún, *Historia general de las cosas de Nueva España*, que en doce libros y dos volumenes escribió, el R. P. fr. Bernardino de Sahagún. Dala a luz con notas y suplementos Carlos María de Bustamante (México: Imp. del ciudadano A. Valdés, 1829–30).

[8]Howard F. Cline, "The *Relaciones Geográficas* of the Spanish Indies, 1577–1586," *Hispanic American Historical Review* 44 (August 1964), 341–374.

[9]For the careers and work of these men see Irving A. Leonard, *op. cit.*; his article "A Great Savant of Colonial Peru; Don Pedro de Peralta," *Philological Quarterly* 12 (January 1933), 54–72; and his book *Don Carlos de Siguenza y Góngora, a Mexican Savant of the Seventeenth Century* (Berkeley: University of California Press, Publications in History 18, 1929).

[10]Lanning, *op. cit.*, Chapters 1–3.

[11]For an account of La Condamine's scientific work, see Edward J. Goodman, *The Explorers of South America* (New York: The Macmillan Company, 1972), pp. 183–203.

[12]*Ibid.*, pp. 245–263. For accounts of later scientific expeditions to Latin America by foreign scientists see Victor Wolfgang Von Hägen, *South America Called Them; Explorations of the Great Naturalists: La Condamine, Humboldt, Darwin, Spruce* (New York: Alfred A. Knopf, 1945), pp. 3–160, and his *The Green World of the Naturalists: A Treasury of Five Centuries of Natural History in South America* (New York: Greenberg Publishers, 1948); and Paul R. Cutright, *The Great Naturalists Explore South America* (New York: The Macmillan Company, 1940). The first English edition of Humboldt's travels was published between 1814 and 1829 as his *Personal Narrative* in 7 volumes.

[13]Iris Higbie Wilson, "Scientists in New Spain: The Eighteenth Century Expeditions," *Journal of the West* 1 (July/October 1962), 24–44.

[14]George Basalla, *op. cit.*, 613.

[15]Stanley J. Stein and Barbara H. Stein, *The Colonial Heritage of Latin America; Essays on Economic Dependence in Perspective* (New York: Oxford University Press, 1970), Chapter 6.

[16]Fernando de Azevedo, *As ciências no Brasil*, 2 vols. (Rio de Janeiro: Edições Melhoramentos, 1955), Vol. 1, pp. 84–88.

[17]For an account of astronomy in South America, see Florian Cajori, *The Early Mathematical Sciences in North America and South America* (Boston: R. G. Badger, 1928).

[18]Charles R. Boxer, *The Golden Age of Brazil, 1695–1750: Growing Pains of a Colonial Society* (Berkeley and Los Angeles: University of California Press, 1964), pp. 30–60.

[19]For a good account of scientific exploration in Brazil in this period see Rodolpho

García, "História das explorações scientíficas," in Instituto Histórico e Geográphico Brasileiro, *Diccionário histórico, geográphico e ethnográphico do Brasil,* 2 vols. (Rio de Janeiro: 1922), Vol. 1, pp. 856-910.

[20]For a history of Portuguese policy towards Brazil, see Charles R. Boxer, *The Portuguese Seaborne Empire 1415-1825* (New York: Alfred A. Knopf, 1969). See pp. 340-366 for a description of education in Brazil in the eighteenth century.

[21]Francisco Guerra, in his *Bibliografia médica brasileira, período colonial 1808-1821* (New Haven, Connecticut: Yale University School of Medicine, 1958), p. 6, describes the founding of a printing press in Rio in 1747, from which three imprints appeared before the Royal House of Portugal ordered its closing.

[22]Charles R. Boxer, *The Golden Age of Brazil*, p. 293.

[23]See Alexander Marchant, "Aspects of the Enlightenment in Brazil," in Arthur P. Whitaker, *op. cit.*, pp. 95-118. For an account of another aspect of the Enlightenment in Brazil, see E. Bradford Burns, "The Role of Azeredo Coutinho in the Enlightenment of Brazil," *Hispanic American Historical Review* 44 (May 1964), 145-160.

[24]C. H. Haring, *Empire in Brazil. A New World Experiment with Monarchy* (Cambridge, Massachusetts: Harvard University Press, 1958), pp. 5-7.

[25]Joseph F. X. Sigaud, in his *Du climat et des maladies du Brésil; ou Statistique médicale de cet empire* (Paris: Fortin, Masson, 1844). Part 4, Chapter II gives an interesting account of medical developments.

[26]See J. P. Fontenelle, "Hygiene e saúde pública," in IHGB, *Diccionário histórico, geográphico e ethnográphico do Brasil*, Vol. 1, pp. 418-463.

[27]Fernando de Azevedo, *Brazilian Culture: An Introduction to the Study of Culture in Brazil.* Translated by W. R. Crawford (New York: The Macmillan Company, 1950), p. 174.

[28]The history of the Military Academy as well as that of other institutions of science in the nineteenth century is covered in some detail in Fernando de Azevedo (*ed.*), *As ciências no Brasil.* See also Umberto Peregrino, *História e projeção das instituições culturais do exército* (Rio de Janeiro: José Olympio Editôra, 1967).

[29]The classic statement of the natural history tradition is found in William M. and Mabel S. C. Smallwood, *Natural History and the American Mind* (New York: Columbia University Press, 1941). This view of the dominance of natural history over the physical sciences is challenged by George H. Daniels in his book, *American Science in the Age of Jackson* (New York: Columbia University Press, 1968), Chapters I and II.

[30]The effect of new systems of classification on the older generation of botanists is well described by A. Hunter Dupree in his *Asa Gray, 1810-1888* (New York: Atheneum, 1968), Chapter II. The rise of specialization and professionalization is discussed in George H. Daniels, "The Process of Professionalization in American Science: The Emergent Period, 1820-1860," *Isis* 58 (Summer 1967), 151-166.

[31]George H. Daniels, in "The Process of Professionalization in American Science: The Emergent Period, 1820-1860," *op. cit.*, shows how, for instance, the usefulness of natural history was emphasized by scientists as a method of ensuring public support for their work.

[32]There are numerous accounts of foreign travelers and naturalists in Brazil. See Fernando de Azevedo's *As ciências no Brasil*, and Candido de Mello Leitão, *História das expedições científicas no Brasil* (São Paulo: Companhia Editôra Nacional, Brasiliana Vol. 209, 1941). The Russian scientific mission is described in G. G. Manizer, *A expedição do acadêmico G. I. Langsdorff ao Brasil (1821-1828)* (São Paulo: Companhia Editôra Nacional, Brasiliana Vol. 329, 1967).

[33]Anyda Marchant, "Dom João's Botanical Garden," *Hispanic American Historical Review* 41 (May 1961), 259-274.

[34]The history of the Museum has been drawn from a number of sources, including the relevant sections in Fernando de Azevedo, *As ciências no Brasil*, and Ladisláu de Souza Mello e Netto's *Le Muséum National de Rio-de Janeiro et son influence sur les sciences naturelles au Brésil* (Paris: Librairie Ch. Delagrave, 1889). See also João Batista de Lacerda, *Fastos do Museu Nacional do Rio de Janeiro: Recordações históricas e scientíficas fundadas em documentos authênticos e informações verídicas.* Obra executada por indicação e sob o patronato do Sr. Ministro do Interior, Dr. J. J. Seabra (Rio de Janeiro: Imprensa Nacional, 1905).

[35]C. de Mello Leitão, *A biologia no Brasil* (São Paulo: Companhia Editôra Nacional, Brasiliana Vol. 99, 1937), p. 177. Mello Leitão states that, once created, the Museum was left in an almost abandoned state for the next one hundred and twenty years. See pp. 169-170.

[36]*Ibid.*, pp. 204-207.

[37]Fritz Müller (1822-1897) was a German naturalist who came to Brazil after the Revolution of 1848, and settled in the German colony of Santa Catarina in 1852. There he spent the next forty-five years of his life, publishing widely in foreign and Brazilian journals in many fields of natural history. He made many striking contributions to the theory of evolution, such as his work on mimicry and "convergent" evolution. His *Für Darwin*, which appeared in 1864, was a defense of Darwin's theory of evolution and was widely read. It appeared in English translation as *Facts and Arguments for Darwin With Additions by the Author,* translated from the German, by W. S. Dallas (London: J. Murray, 1869). A description of Müller's life and work appears in Fernando de Azevedo, *As ciências no Brasil*, Vol. 2, pp. 207-211. Another account, and translations from the Portuguese and German of some of Müller's papers, appear in George B. Longstaff, *Butterfly-Hunting in Many Lands; Notes of a Field Naturalist* (London: Longmans, Green, and Co., 1912), pp. 601-666.

The naturalist and paleontologist, Peter Wilhelm Lund (1801-1880) made many contributions to the understanding of Brazilian fossils during his stay in the state of Minas Gerais. He was also the first to publish a scientific paper on a New World human skull found in association with extinct animals and to thereby raise the possibility of a great antiquity for man in the New World. See Kenneth MacGowan and Joseph A. Hester, Jr., *Early Man in the New World* (Garden City, New York: Anchor Books, Doubleday and Company, 1962), p. 121. For an account of his work in general, see Anibal Mattos, *O sábio Dr. Lund e estudos sôbre a pre-história brasileira. Vida e obra de Peter Wilhelm Lund, ethnographia, archeologia, anthropologia, antiguidade do homem no Brasil* (Bello Horizonte: Edições Apóllo, 1935).

The originality and quality of the work of these scientists is striking; trained in a formative period of the geological, paleontological, and biological sciences, even their geographical and intellectual isolation did not deter them from making interesting discoveries.

[38]The following are population figures for the federal district of Rio de Janeiro calculated by the Department of Public Health in Rio in 1910: 1821, 112, 695; 1833, 137, 078; 1870, 235, 381; 1890, 522, 651; and 1906, 811, 443. See Brazil, Directoria Geral de Saúde Pública, *Annuário de estatística demographo-sanitária 1910* (Rio de Janeiro: Imprensa Nacional, 1913), p. 14. According to the same source, by 1907 São Paulo had grown to a population of 300,000, San Salvador (Bahia) to 265,000, and Recife to 186,000. See Brazil, Directorial Geral de Saúde Pública, *Annuário de estatística demographo-sanitária, 1907* (Rio de Janeiro: Imprensa Nacional, 1909), p. 71.

[39]J. Fred Rippy, *Latin America and the Industrial Age* (New York: G. P. Putnam's Sons, 1944), pp. 26–27, shows that by 1863 only 180 miles of railroad had been completed.

[40]Domingos José Freire's opinion on the plan to create a university in Brazil, and his proposals for the reform of medical education in Brazil, are reprinted in Francisco Bruno Lobo, *Uma universidade no Rio de Janeiro*, 2 vols. (Rio de Janeiro: Oficina Gráfica da Universidade Federal do Rio de Janeiro, 1967), Vol 1, pp. 251–255.

[41]"The Present State of Science in Brazil," *Science* 1 (March 1883), 211–214. The article appeared anonymously. Alpheu Diniz Gonsalves, in his bibliography of Derby's works, attributes this article to Derby; this attribution is almost certainly correct. See Alpheu Diniz Gonsalves (*ed.*), *Orville Derby's Studies on the Paleontology of Brazil: Selection and Co-ordination of This Geologist's Out of Print and Rare Works* (Rio de Janeiro: Published under the Direction of the Executive Commission for the First Centenary Commemorating the Birth of Orville A. Derby, 1952.)

[42]Louis Agassiz and Elizabeth C. Agassiz, *A Journey in Brazil* (Boston: Ticknor and Fields, 1868), p. 499. In a similar vein, the French engineer Louis Léger Vauthier, who visited Brazil between 1840 and 1846 as chief of a French scientific mission in the state of Pernambuco to study the construction of railroads wrote in his diary:

"I should like to see some one ask of these idlers who talk so much about patriotism to carry a surveyor's level and help take a level. He would say right away that he is not a *servant* or a slave, that he is a *freeman* and was not born to carry things." Quoted in Fernando de Azevedo, *Brazilian Culture: An Introduction to the Study of Culture in Brazil*, p. 174. This diary was written in French and later published in Portuguese under Gilberto Freyre's initiative as *Diário íntimo do engenheiro Vauthier, 1840–1846* (Rio de Janeiro: Serviço Gráfico do Ministério da Educação e Saúde, 1940).

[43]Louis Agassiz and Elizabeth C. Agassiz, *op. cit.*, pp. 499–500.

[44]*Ibid.*, pp. 501 and 502. In a more general vein Agassiz commented, "Even now, after half a century of independent existence, intellectual progress in Brazil is manifested rather as a tendency, a desire, so to speak, giving a progressive movement to society, than as a positive fact. The intellectual life of the nation when fully developed, has its material existence in large and various institutions of learning, scattered throughout the country. Except in a very limited and local sense, this is not yet the case in Brazil." See p. 498.

[45]Details about Frederick Hartt's career in Brazil are found in Richard Rathbun, *Sketch*

of the Life and Scientific Work of Professor C. F. Hartt. Read Before the Boston Society of Natural History (Boston, 1878), and in Fernando de Azevedo, *As ciências no Brasil*, Vol. 1, pp. 250–252. Hartt's textbook of Brazilian geology appeared as *Geology and Physical Geography of Brazil* (Boston: Fields, Osgood and Co., 1870).

[46]Ladisláu de Souza Mello e Netto, *op. cit.*, p. 7. Ladisláu signed himself Ladisláu Netto.

[47]This was his *Investigações históricas e scientíficas sôbre o Museu Nacional do Rio-de-Janeiro*, referred to in Netto's *Le Muséum National de Rio-de-Janeiro*, p. 8.

[48]Hermann von Ihering was a well-known zoologist and authority on Brazilian molluscs. He directed the Museu Paulista from 1894 to 1915. He authored *The Anthropology of the State of São Paulo, Brazil* (São Paulo: Duprat and Co., 1904), as well as numerous articles in Portuguese and German.

Louis Couty had originally come to Brazil in 1876 on contract to the Brazilian government as part of the general governmental effort to improve the sciences in Brazil. In an interesting evaluation of science in Brazil, he characterized the impediments to the development of science as "bad scientific administration, and the difficulty of finding students." He also described how research funds were cut to make up deficits in the budget elsewhere. See Louis Couty, "O ensino superior no Brasil," *Gazeta Médica da Bahia* 15 (Maio 1884), 521–532.

[49]A guide to the Brazilian exhibition of 1882 was published by the Museum. See Rio de Janeiro, Museu Nacional, *Guia da exposição anthropológica brasileira realizada pelo Museu Nacional do Rio de Janeiro, à 29 de julho de 1882* (Rio de Janeiro: G. Leuzinger e Filhos, 1882).

[50]Louis Couty's work in physiology is described in Rio de Janeiro, Museu Nacional, *João Batista de Lacerda. Comemoração do centenário de nascimento, 1846–1946* (Rio de Janeiro: Museu Nacional, Publicações Avulsas, No. 6, 1951). Louis Couty published an outline for a course in experimental biology in 1880, in which he wrote: "Vous savez aussi bien que moi . . . au Brésil comme en France autrefois, il est le mode dans un certain nuclei de n'attacher aucune importance aux recherches du laboratoire: on dit volontiers qu'elles aboutissent toujours à des spéculations vagues ou à des conclusions scientifiques sans utilité pratique . . . " See Louis Couty, Musée National, *Cours de biologie expérimentale. Leçon d'ouverture* (Rio de Janeiro: G. Leuzinger e Fils, 1880).

Lacerda's contributions are evaluated in Rio de Janeiro, Museu Nacional, *João Batista de Lacerda. Comemoração do centenário de nascimento, 1846–1946.* This publication also contains a bibliography of his work. Another account of Lacerda and Couty's contributions to physiology is found in Fernando de Azevedo. *As ciências no Brasil*, Vol. 2, pp. 215–221.

[51]Francisco Bruno Lobo (*ed.*), *Membória histbórica dos accontecimentos mais notáveis occorridos na Faculdade de Medicina do Rio de Janeiro em 1879*, redigida pelo Dr. Nuno de Andrade, lente substituto da secção médica. Mimeographed.

[52]Francisco Bruno Lobo (*ed.*), *Relatório do director da Faculdade de Medicina do Rio de Janeiro por 1883*, pelo Visconde de Saboia, p. 52. Mimeographed.

[53]In addition to Derby, Hartt stimulated the interest of Richard Rathbun and John Casper Branner in Brazilian geology.

[54]See Thomas G. Manning, *Government in Science: The U.S. Geological Survey,*

1867-1894 (Lexington: University of Kentucky Press, 1967), Chapter 4.

[55]Derby's life and career are described in John Casper Branner, "Memorial of Orville A. Derby," *Bulletin of the Geological Society of America* 27 (1916), 15-21. A bibliography of Derby's work up until 1903 appears in Branner's *A Bibliography of the Geology, Mineralogy and Paleontology of Brazil* (Rio de Janeiro: Imprensa Nacional, 1903). Another bibliography appears in Alpheu Diniz Gonsalves, *op. cit.*

[56]Poppino characterizes the period of industrial growth in Brazil between 1870 and 1920 as one of dependence on foreign capital and technology. See Rollie E. Poppino, *Brazil, The Land and People* (New York: Oxford University Press, 1968), pp. 200-237.

[57]Victor Leinz, "A geologia e a paleontologia no Brasil," in Fernando de Azęvedo, *As ciências no Brasil*, Vol. 1, p. 252. Although not the first to study Brazilian geological formations, Leinz calls him the last of the great foreign and the first Brazilian geologist.

[58]For a discussion of Agassiz' marriage and his position in American science see the biography by Edward Lurie, *Louis Agassiz, A Life in Science* (Chicago: University of Chicago Press, 1960).

[59]In 1907 a new federal survey was organized and Derby appointed Director. Derby committed suicide in 1915, shortly after naturalization as a Brazilian citizen; his death was attributed to his disappointment when the survey was suddenly cut back.

[60]To borrow from the language of economics, one might say that foreign scientists create a smaller multiplier effect and weaker backward linkages than a national scientist in forming scientific institutions that can survive without continued foreign aid. On the other hand, the original stimulus to the development of a particular branch of science very often comes from a foreign scientist who, by accident or design, finds himself working in a developing country and encourages nationals to work in his field of interest. The impact will depend on many factors, such as where the foreigner comes from, the area of science in which he is interested, and the degree of government interest. Migrations of large numbers of scientists are rare, and such groups may have an unprecedented effect on the intellectual life of the nation. The impact of European intellectuals fleeing fascism in the 1920s and 1930s on American culture is explored in Donald Fleming and Bernard Bailyn, *The Intellectual Migration; Europe and America, 1930-1960* (Cambridge, Massachusetts: The Belknap Press of Harvard University Press, 1969). Scientists fleeing fascist Europe were also invited to Brazil by the São Paulo authorities and similarly had a great impact on Brazilian science. See, for example, James Rowe, "Science and Politics in Brazil: Background of the 1967 Debate on Nuclear Energy Policy," in Kalman Silvert (*ed.*), *The Social Reality of Scientific Myth, Science and Change* (New York: American Field Staff, Inc., 1969), pp. 91-122.

[61]George Basalla, "The Spread of Western Science," *op. cit.*, 617-620.

[62]See Chapter 8, section *National Capabilities in Technology*, for a discussion of this issue.

[63]Joseph Ben-David, *The Scientist's Role in Society. A Comparative Study* (Englewood Cliffs, New Jersey: Prentice-Hall, Inc., Foundations of Modern Sociology Series, 1971), pp. 123-129, 142-162.

[64]*Ibid.*, pp. 108-109.

3

Medicine in Brazil: Nineteenth Century Background

Standing somewhat distinct from the tradition of science described in the previous chapter was the medical and public health tradition in Brazil. While institutionally the field of medical science came to share many of the characteristics described in the case of the other sciences, the urgent need for physicians when the Portuguese court came to Brazil in 1808, and the severity of epidemic disease, helped give medicine a professional identity and social visibility long before the social role of scientists was clearly defined.

In this respect, Brazil was typical of new nations in the evolution of its indigenous scientific establishment. Physicians usually form the first, and the largest, group of professionally trained personnel in new countries. Medical institutions, including medical schools, hospitals, and medical societies, are usually among the first institutions of science to be founded, since the need for medical care always exists. As a social profession, the practice of medicine is shaped by the values of a society and varies from country to country. Medicine is also shaped by developments in the biomedical sciences, while the applied nature of medicine provides opportunities for the creation of indigenous capabilities in medical science. The fact that Oswaldo Cruz was a medical doctor, and that the first research tradition to be established in Brazil lay in the biomedical field, justifies a separate consideration of the medical tradition in Brazil.[1]

Until the transferral of the court to Brazil in 1808, this medical tradition was shaped by the scarcity of doctors, the absence of training facilities in the country, and above all by the high incidence of epidemic diseases whose causes were unknown to medical scientists until the end of the nineteenth century. The first visitors and settlers in Brazil were favorably impressed with the beauty of the tropical jungles and the pleasant climate. But as colonization proceeded and slaves were introduced into the colonial economy, the incidence of epidemic and non-epidemic disease grew rapidly. By the end of the nineteenth century the vision of Brazil as a tropical paradise had long disappeared, and the climate had been established in most people's minds as the chief cause of disease as well as the main impediment to the emergence of civilization in the country. The Oswaldo Cruz Institute's ability to substitute a microbiological explanation of disease for the traditional climatological one was one of the reasons for Cruz' prestige within Brazil in the decade preceeding World War I.

Many of the epidemic and non-epidemic diseases encountered in Brazil were already well known to Europeans—diseases such as malaria, smallpox, dysentery, measles, and tuberculosis. Others, such as various tropical ulcers and yellow fever, were new. Yellow fever was particularly important to Brazil's history. Many epidemics occurred in the sixteenth and seventeenth centuries, and the first published description of an epidemic to appear anywhere was published in Lisbon by a Brazilian. This work, the *Tratado unico da constituição pestilencial de Pernambuco* by João Ferreira da Rosa, described the epidemic of yellow fever in northeastern Brazil between 1680 and 1694, an epidemic which killed thousands.[2] After the seventeenth century yellow fever did not disappear from Brazil, but was confined to local outbreaks. It returned in epidemic proportions in 1849, and thereafter had a profound, though as yet inadequately assessed, effect on the economy and social life of the country. Its role in stimulating science in Brazil is described in Chapters 4 and 5.

Dealing with yellow fever epidemics strained the resources and organizational abilities of even the most advanced nations throughout the nineteenth century. Diagnosis itself was hard, yellow fever often being confused with malarial fever so that medical statistics of the past do not always give a reliable picture of yellow fever's incidence. Attacks, which were short in duration and resulted in death or recovery in five to seven days, were often accompanied in the later phase of the illness by the infamous "black vomit." The mortality rate was high, particularly for Europeans, and death often occurred with convulsions or coma.

The origin and cause of the disease gave rise to much speculation throughout the eighteenth and nineteenth centuries. There were differences of opinion as to whether the disease was contagious by direct contact or not, or through contact with infected clothing, or "fomites," as they were called.[3] The question of local versus external origin was also debated heatedly. The disease sometimes seemed to travel long distances over water and appear when ships arrived in port.

At other times it appeared to flare up suddenly in a localized area and remain in that area alone. Because outbreaks of yellow fever often occurred with the arrival of European ships from cold climates, the hot climate of Brazil was believed to be one cause of the disease. Some believed its cause was a "miasm," or poison, activated by the heat, under conditions of crowding or dirt, which were also associated with epidemics. Others thought the "virus" lay in the soil, and was released when the soil was disturbed, as in the construction of roads and houses. What was not understood until 1900 was that yellow fever is transmitted from person to person via a specific mosquito, and that childhood contact with the disease confers immunity, so that native populations are relatively protected, while arriving Europeans face a high risk. Yellow fever inspired horror, and it was this disease, together with malaria, that made Europeans fear the tropics.

The main stimulus to medical study in Brazil before 1808 was the desire to understand disease and to discover plants with medicinal properties.[4] A number of books written by Brazilians and visitors to Brazil were published in the seventeenth and eighteenth centuries in Portugal and France. But few physicians came to Brazil to study disease in any systematic way, or to practice their profession.[5] The vastness of the interior, the sparseness of the colonial population and its dispersion, did little to attract doctors. Unlike the Spanish American colonies, Brazil possessed no medical schools before the nineteenth century, although individual surgeons and physicians occasionally offered lectures in Bahia or Rio de Janeiro by the end of the eighteenth century, and the hospitals organized by the Brotherhood of the Misericordia produced practical surgeons. Few licensed physicians with university degrees practiced in the colony. Ostensibly medicine and pharmacy were regulated by the Surgeon-General of the Army. In fact, local druggists, faith healers, herbalists and plantation owners looked after the medical needs of the population as best as they could. Given the lack of scientific knowledge of the causes of most of the most serious diseases of the times, such as dysentery, tuberculosis, or measles, the results arrived at by lay treatment were probably not very different from those arrived at by licensed physicians, and possibly better when the worst excesses of bleeding and purging were avoided. The hospitals of the Santa Casas de Misericordia, though often without attending physicians, at least provided some beds and food for the sick poor.[6]

Medicine in the Nineteenth Century

The professional organization and regulation of medicine in Brazil began only in the nineteenth century, and led to a strong tradition of clinical medicine in two cities, Rio de Janeiro and Bahia (Salvador). Outside the cities, there was virtually no organization of private or public health, and the majority of Brazilians lived in conditions of poverty and disease. Three different aspects of the medical tradition need to be considered: the evolution of medical education, the evolution of public health legislation, and the evolution of medical research. All three traditions came together in the Oswaldo Cruz Institute to create the first productive work in medical science in Brazil in the early twentieth century.

Formal medical education started in Brazil in 1808 when physicians and surgeons were authorized by the Crown to offer courses in surgery and anatomy in the military hospitals of Bahia and Rio de Janeiro. New chairs were slowly added, the courses formalized as Academies, and eventually in 1832 two permanent, degree-granting Medical Schools were established. The directors of the two schools were nominated by the government from lists drawn up by the professors of each school. The professors themselves were selected through a process of competitive examination, and were paid a salary for their teaching services.

The high academic standards required for entrance into the schools, and the fact that only two schools existed in the entire country until the early twentieth century, severely limited the number of doctors graduating. Here the situation differed greatly from that in the United States, where proprietorial schools of medicine proliferated. The French physician Joseph F. X. Sigaud, whose book *Du climat et des maladies du Brésil; ou Statistique médicale de cet empire* (1844) is not only a classic of epidemiology but also a valuable source of information about medical practice in Brazil, reported that in the ten years between 1833 and 1843, out of a total of two thousand students entering the Medical School in Rio de Janeiro only one hundred emerged as licensed physicians.[7] The rest either left voluntarily or failed their examinations. The annual reports of the Medical School in Rio demonstrate a slow rise in numbers, until by 1882 the total number of students attending in all years of the curriculum was over one thousand five hundred.[8] However, as the report in 1879 made clear, a large percentage of those graduating never practiced, the degree being sought for social purposes rather than as a means to professional practice.[9] This was partly because, until universities were founded in the twentieth century, law and medicine, with the exception of civil engineering after 1854, were the only professional degrees available to the Brazilian student.

Though their numbers were small, Brazilian physicians enjoyed a high social status and were genuinely concerned with professional standards.[10] There was a lively interest in epidemic disease, and Sigaud's book was one outstanding manifestation of this interest.[11] The influential Society of Medicine (Sociedade de Medicina) founded in 1829 by four Brazilian and four foreign physicians (of whom Sigaud was one), played an important role in advising the national government during medical crises.[12] Medical journals, such as the *Propagador das sciencias médicas*, published by Sigaud, and the *Seminário de sáude pública*, provided medical information for the interested. When epidemics broke out in and around Rio, the Society and the Medical School sent their members and medical students into epidemic areas to study the disease and treat the sick free of charge.

Ignorance about the causes of disease and the mechanisms of transmission limited doctors everywhere from dealing adequately with illness. Treatment in Brazil, as elsewhere in the world, was based on the few specific remedies tested by time, such as vaccination for smallpox (introduced early in the century in Rio and Bahia), digitalis for the treatment of heart disease, the use of quinine, often in huge doses, for malaria and intermittent fevers, and mercury compounds for venereal diseases. Bleeding and purging were other traditional standbys. Patent medicines, despised by the regular physicians, were widely used, especially in the countryside where physicians were rare. An interesting account of the medical treatment received by the explorer and traveler Richard Burton in 1868, while consul in Santos, has been left us by his wife Isabel. Burton was struck by what was probably an acute crisis brought on by a lung infection and chronic alcoholism. His wife wrote to her mother that:

> Richard was very ill with a pain in the side. At last he took to incessant paroxysms of screaming and seemed to be dying, and I knew not what do do. Fortunately a doctor came from Rio on the eighth day of his illness . . . He put twelve leeches on, and cupped

> him on the right breast, lanced him in thirty-eight places, and put on a powerful blister on the whole of that side. He lost an immense deal of black clotted blood . . . The agony was fearful, and poor Richard could not move hand or foot, nor speak, swallow, or breathe without a paroxysm of pain that made him scream for a quarter of an hour. When I thought he was dying, I took the scapulars and some holy water, and I said, "The doctor has tried all his remedies; now let me try one of mine." I put some holy water on his head and knelt down and said some prayers, and put on the blessed scapulars . . . He was quite still for about an hour, and then he said in a whisper, "Zoo, I think I'm a little better."[13]

Of the treatment she added, "there was something to be given or rubbed every half-hour, of which a very large ingredient was orange tea."[14]

Public health legislation had a long and complicated history in Brazil which cannot be dealt with here in any detail, but, once again, inadequate medical knowledge about the origin of disease and the failure to impose quarantines and other sanitary measures prevented most of the sanitary laws from being effective during epidemics.[15] The economic and social costs of epidemic disease cannot be calculated, partly owing to the absence of reliable vital statistics. Within Rio de Janeiro city the national and municipal authorities, whose responsibilities for sanitation overlapped in many areas, combined with physicians from the Medical School and members of the Society of Medicine of Rio to deal with epidemics when they arose. In 1829, for instance, a severe outbreak of intermittent fever broke out in the town of Macacú, outside the capital city in the province of Rio. The French physician Sigaud described in some detail how the authorities and the doctors responded to the crisis. Macacú was situated on the side of a river, in an area rich in forests, large plantations of coffee, and fields of rice and manioc. The season had begun with a very dry spell, which had then been followed with heavy rains. When the epidemic broke out, the national government nominated some doctors to form a commission to go to Macacú to study the disease. They were aided by another commission of doctors sponsored by the Society of Medicine in Rio, which also held a series of special sessions to discuss questions of hygiene and therapy. Many of the sick came to Rio for treatment and help, and some effort was made to segregate the sick from the well.

Doctors in Rio treated the disease with large doses of quinine sulphate, which brought on nervous attacks, and with "Leroy" treatment, a French purgative which caused severe vomiting. Many doctors went voluntarily to Macacú to treat the ill, and several autopsies were also carried out to discover the cause of deaths.[16]

Outside the large cities, disease was rampant, medicine and hygiene were very simple, and licensed doctors were consulted only in extreme cases. Stein describes the conditions of the slaves in the coffee county of Vassouras in the 1850s and 1860s and gives a good picture of the medical conditions and treatment of the times. Slaves were often poorly fed, poorly clothed, and worked extremely hard. They were not expected to live long, and a good proportion of the slave work force was sick at any one time, common illnesses being *pulex penetrans* (*bicho de pé* and now called *Tunga penetrans* which caused ulcers on the foot preventing work), erysipelas, tuberculosis, and diarrheas. When a slave fell sick, he would be treated with home remedies or patent medicines, often administered by the plantation owner who had no real understanding of what he was doing. Or a *curandeiro* was called in, a person with a local reputation for healing who relied on herbal remedies and other home cures.[17]

Turning now to medical research, for most of the nineteenth century, such research did not form a part of the average physician's work any more than it did for the physician in Europe or the United States.[18] However, medical knowledge greatly improved during the second half of the century and eventually brought about a new emphasis on research. The public health movement in Europe and the United States was spurred by the terrible waves of cholera which spread across country after country starting in the winter of 1831–1832; slowly an understanding of its contagious nature and the relation between epidemic disease and dirt and poverty was gained. Government boards of health were established to regulate public sanitation of streets and houses and impose quarantines.[19] The development of bacteriology also gave an understanding of the microbic origin of many diseases. The role of the insect vector was worked out for malaria by 1894. Serum therapy and immunization were added to the arsenal of medical weapons against disease.

The Medical Schools of Bahia and Rio de Janeiro responded to these changes by improvements in both the physical conditions of the schools as well as the faculties. In Bahia, the Brazilian-born, German-educated physician Otto Wucherer (1820–1873), son of German and Dutch parents, the Englishman John Patterson (1820–1882), and the Brazilians José Francisco da Silva Lima and Silva Araújo initiated a decade of experimental pathology, beginning with Wucherer's identification of filariasis as a specific disease rather than a general anemia, and his discovery of the larval stage of the filariae. His colleague, Patterson, was among the first to diagnose yellow fever when it returned to Brazil in 1849 in epidemic proportions, and was also responsible for founding the *Gazeta Médica da Bahia*, a leading medical journal in the second half of the century. None of these experimentalists held chairs in the Bahia Medical School, however, and this fact restricted the training of students in experimental pathology and prevented continuation of the tradition of research.[20]

In Rio de Janeiro, the teaching doctors led the way in urging the government to improve the curriculum and facilities in their Medical School. They were hamstrung in their efforts by the rigid centralization of the bureaucracy and its slowness to respond to their requests. The school had been plagued for many years by inadequate clinical and laboratory facilities, unruly students, and constant but ineffective changes in regulations as well as in its location.[21] The annual reports made up a litany of woes, stretching from broken and incomplete equipment, bad teaching, absent professors, and absent students. In 1876, the author of the report concluded that the buildings in which the faculty worked lacked "all the requirements for the proper conduct of teaching—order, respect, cleanliness. For eighteen professors there are only three rooms, each one worse than the next."[22] Three years later, the hygienist Dr. Nuno de Andrade brought matters to a head with a savage indictment of government inaction.[23] He referred in the report to France's humiliating defeat at the hands of Prussia, and drew attention to the relation between science and national power, a relation which Pasteur himself had stressed in an article on the role of the government in science.[24] Nuno de Andrade pointed to the need to improve the scientific base of medicine and the school to be free from restrictive government regulations. He even went so far as to suggest that women should be allowed to attend medical school.

The government finally took action in 1880 when a number of new laboratories were constructed. At the same time an attempt was made to allow medical students to pursue their own interests and free the teaching staff from rigid schedules by making attendance once more a matter of choice, rather than compulsion. One result, however, of this reform was that class attendance fell off and practical examinations in laboratory medicine became a farce.

The deficiencies in teaching in laboratory medicine were important because the Medical Schools of Rio and Bahia were the sole source of trained medical scientists inside Brazil. The strong clinical tradition tended to draw away students from an interest in experimentation. In 1901 a new law established a chair in microbiology and re-established compulsory attendance in laboratory classes, but though the number of medical students was increasing, the quality of teaching of scientific medicine did not improve rapidly. In 1904, for instance, the Minister of Justice and the Interior, under whose jurisdiction the Medical School lay, noted in his annual report to Congress that the professor of microbiology was attempting to instruct one hundred and fifty students with a single microscope. The situation in histology and pathology was much the same.[25] Clearly the laboratory approach to medicine, and even the use of the microscope, were not routine parts of medical education in Brazil in 1900.

It must not be thought that Brazil was in this respect very different from the United States. The Flexner Report of 1910 exposed the fact that in a very large number of schools in America the standards of medical education were abysmally low.[26] The difference between Brazil and the United States lay in the fact that in America a few independently created schools and institutes had begun to introduce the new methods of bacteriology and pathology. The Johns Hopkins Medical School in particular pioneered changes in medicine and eventually stimulated reforms in medical education in the United States. Although no new schools of medicine were founded in this period in Brazil to introduce scientific medicine into the country, the traditionalism of the Rio Medical School was such as to suggest that any future developments in bacteriology or pathology would in fact take place outside of the Medical School, in independent research-oriented institutions. This was mainly because, institutionally at any rate, it was often easier to set up totally new solutions to problems of scientific administration than to attempt to reform existing institutions.

The history of the Rio Medical School is interesting from another point of view, namely the effect of the physical construction of laboratories on the advance of science. The experience of Brazil suggests that construction of laboratories *per se* cannot guarantee the success of experimental science. A solution to the problems of government support for science, student recruitment, and of leadership in science were of greater significance. The French industrial biologist, Louis Couty, made a similar evaluation of the factors most important for the advance of science in a developing country, when he commented on medical education in Brazil in 1884.[27] Though he believed the laboratories in Brazil were generally of better quality than those in the other countries of South America, he argued that the existence of laboratories, botanical gardens and museums were not *essential* for outstanding scientific work. The main difficulties science faced in Brazil, he believed, were rather those of finding students of science, the small number of scientists as a whole, and poor scientific administration. *A propos* of this latter factor, he mentioned the lack of budgetary autonomy of administrators of scientific institutions in Brazil.

Medicine in the 1900s

Despite the institutional limitations to experimental medicine, there were some signs by 1900 that the situation in medicine and public health was ripe for improvement. Immigration into the cities of Rio de Janeiro and São Paulo had swollen the urban population, exacerbating health conditions while at the same time introducing illnesses hitherto still somewhat rare in the country, such as cholera. The threat epidemics posed to the valuable labor force was a subject of concern to coffee plantation owners. Engineers and doctors rose to positions of prominence, especially in the city of São Paulo, which was rapidly emerging as one of the most important and forward-looking of Brazil's cities.[28] New intellectual movements, such as Darwinism and positivism, had an indirect influence on science by making the intellectual environment in Brazil more hospitable to scientific ideas. The Paulista physician and positivist Luís Pereira Barreto, for instance, was at the forefront of the sanitation and public health movement in the state of São Paulo.[29]

Economically and politically, the year 1900 also marked something of a turning point. The first years of the Republic had been stormy and difficult. Balance of payment difficulties had plagued the country, until a financial settlement with England was negotiated and an export boom begun that gave the government a financial prosperity unequaled in the preceding years of the Republic. Politically, the first years of the twentieth century were years of political optimism. The new President Rodrigues Alves, elected in 1902, had campaigned on the theme of the need for a renewal in Brazil's cultural, social and political life.[30] The political optimism of these years coincided with the rise of interest in the microbiological sciences and the growing conviction among a few intellectuals that science could come to the aid of developing countries in the race toward progress.

In the Brazilian context, confidence about Brazil's future status as a world power and the role that science could play in bringing it about was not only historically a new phenomenon, but in retrospect was particularly important when viewed in the light of the country's historic pessimism concerning its racial and social destiny. Traditionally, when climatological explanations of the causes of disease were accepted by doctors, Brazil's tropical climate was blamed for the diseases endemic and epidemic in the country. It was also assumed that Brazil's racially mixed population was sensual and passive, susceptible to disease, and incapable of the individual or collective control and rationality required for the advance of civilization.[31] The belief that Brazil was doomed to fall behind in the race toward progress had been recently reinforced by the imperialist adventures of the United States in the Carribbean in 1898, adventures based to some extent on the belief that the United States brought to South America a technological, moral, and political superiority that was not found in either the traditional colonizing countries of Spain or Portugal, nor in the racially mixed populations of South America. Some anthropologists publicly expressed their view that without this technological superiority the white man in the tropics would be overwhelmed by the climate, and that intermarriage between the white man and other races would result in a "degradation" of the white man's civilization. As evidence that nowhere had "halfbreeds" produced a high civilization, these anthropologists pointed to Latin Americans, who it was claimed "were paying for their racial liberality," as would all Europeans who "participated in the unhappy mixture of races which was everywhere the curse of tropical states."[32] The fact that foreigners coming to Brazil were without immunity acquired by childhood exposure to yellow fever, and were therefore particularly susceptible to the disease, only seemed to confirm the validity of the belief that Brazil's racially mixed peoples and tropical climate doomed her to disease and backwardness.

Seen against this traditional racial and medical pessimism, the growing feeling among some Brazilian intellectuals that science could provide effective mechanisms for the control of disease was a very important step towards breaking not only a tradition of passivity in the face of medical disaster, but also a traditional indifference to the institutionalization of science. One of the first to state that the key to Brazil's future lay in science was the positivist physician Pereira Barreto, who was a central figure in the sanitation movement that was helping re-vitalize the city of São Paulo in the 1890s.[33] In a letter to the director of public health in São Paulo, Dr. Emílio Ribas, Pereira Bareto remarked that he ranked the solving of yellow fever as of first importance to the nation, and gave the following reasons for this:

> It involves proclaiming to the world now watching us that yellow fever is not the daughter of climatological agencies that are peculiar to our country but the result of factors affecting all countries and upsetting the life of all nations. I am far from putting patriotism above the claims of scientific truth. It would be unworthy of anyone to prejudice the advance of science of humanity for the sake of benefitting himself. But once it is realized that the situation in our country is a scientific and world-wide one, I believe that we shall be able to unite all our efforts to find an answer to the question and in this way re-establish our credit abroad. If it is true, as I am convinced, that yellow fever is transmitted by water, then Brazil finds herself in excellent company. There is no climate in the world that guarantees immunity from invasion of the disease. And a corollary fact that consoles us is that there is no country now being ruined by the disease that cannot in the future restore itself to health.[34]

Thus through science Brazil could escape an historic fatalism concerning its ability to change events, and the nation could emerge as part of the modern, civilized world.

In 1900, however, despite the advances in sanitation in the state of São Paulo, where Emílio Ribas and Pereira Barreto both worked, these views represented those of a minority. The average intellectual had little knowledge of science and cared little about its advance. Technological resources for large-scale sanitation programs were limited, and experimental medicine barely institutionalized in the medical schools of Rio and Bahia. Public disillusionment over public health programs was considerable. On the other hand, epidemics which might be eliminated by new scientific methods offered, potentially at least, a promising source of support for medicine and public health.

The real key, however, to any long-term development of the medical and public health sciences in Brazil was yellow fever. It was not surprising, given the great damage this disease had done to Brazilian life, that yellow fever was uppermost in Pereira Barreto's mind in writing the letter just quoted concerning the role of science in the advance of Brazil. The disease, which appears to have been restricted to local outbreaks for over two hundred years, returned in epidemic proportions in 1849. Starting in the city of Salvador in Bahia, the disease swept south, infecting more than 90,000 people. Over four thousand died in the city of Rio de Janeiro alone before the first epidemic was over. From 1849 on, yellow fever was endemic along the coast and epidemic in the major cities, the mortality rate being very high.[35]

Meanwhile, the last great epidemic in the United States to affect the northern states occurred in 1878. Thereafter the disease retreated southward, reducing the sense of urgency in the north about finding a solution to the question of its mode of transmission. The southern states of the United States, disorganized and politically weakened by the Civil War and Reconstruction, were not in a position to mount a full-scale attack on the disease.[36] In South America, the first generation of bacteriologists turned their attention to yellow fever in the hope of solving the problem of the origin of this terrible scourge. Unfortunately yellow fever is not a bacterial but a viral disease, and its transmission via infected mosquitoes was still only a hypothesis rejected by the majority in the medical profession.[37] It was not until American troops occupied Havana in 1899, and the threat of a massive epidemic of yellow fever in the city forced the American authorities to pay serious attention to the hypothesis of Dr. Carlos Finlay as a last resort, that the transmission by the *Aedes* mosquito was proven correct.[38]

Until 1900, the regularity with which yellow fever attacked Rio de Janeiro, the great susceptibility of foreigners to the disease, and the failure of medicine to solve the question of its cause gave Brazil the reputation as of one of the most unsalubrious areas in the tropics. Discovery of the cause of yellow fever in 1900, therefore, opened up a new era for medicine in Brazil.

References

[1]Donald Cooper, in "Oswaldo Cruz and the Impact of Yellow Fever on Brazilian History," *The Bulletin of the Tulane University Medical Faculty* 26 (February 1967), 49–52, points out that there is as yet no comprehensive medical history of Brazil, and that perhaps the major gap is the lack of a detailed, reliable account of the history of yellow fever in the country.

[2]According to H. Harold Scott, in *A History of Tropical Medicine*, 2 vols. (Baltimore: The Williams and Wilkins Company, 1942), Vol. 1, pp. 280–282, the term "yellow fever" itself was first employed by Griffith Hughes in 1750 in his *Natural History of Barbados*, in order to distinguish it from bilious remittent or other forms of malaria. Yellow fever was often called merely a "plague," a "pestilential fever," a "contagion," an "epidemic" or a "stupor" (*la modorra*). Scott's long chapter on yellow fever is altogether valuable.

[3]By the end of the nineteenth century, doctors were generally agreed that yellow fever was not contagious by direct contact but was contagious through infected articles, even though experimental disproof of this had been given as early as 1804. See Sigusmund Peller, "Walter Reed, Carlos Finlay, and their Predecessors Around 1800," *Bulletin of the History of Medicine* 33 (1959), 195–211.

[4]The history of medicine and public health in Brazil has been taken from a number of sources, including Renato Clark Bacellar, *Brazil's Contribution to Tropical Medicine and Malaria: Personalities and Institutions.* Translated by Anita Farquhar (Rio de Janeiro: Gráfica Olympica Editôra, 1963), Leonídio Ribeiro, *Brazilian Medical Contributions* (Rio de Janeiro: José Olympio Editôra, 1939), Aristides A. Moll, *Aesculapius in Latin America* (Philadelphia and London: W. B. Saunders Company, 1944), Lycurgo Santos Filho, *História da medicina no Brasil: do século XVI ao século XIX* (São Paulo: Editôra Brasiliense Ltda, 1947), Fernando Magalhães, *O centenário da Faculdade de Medicina do Rio de Janeiro, 1832–1932* (Rio de Janeiro: A. P. Barthel, 1932), Alfredo Nascimento, *O centenário da Academia Nacional de Medicina do Rio de Janeiro, 1829–1929* (Rio de Janeiro: Imprensa Nacional, 1929), and Ernesto de Souza Campos, *Instituições culturais e de educação superior no Brasil: resumo histórico* (Rio de Janeiro: Imprensa Nacional, 1941). Some of the annual reports made by the directors of the Medical School of Rio have recently been published by Francisco Bruno Lobo, in his *O ensino da medicina no Rio de Janeiro*, 2 vols. (Rio de Janeiro: Oficina Gráfica da Universidade do Brasil, 1964). Professor Bruno Lobo very kindly allowed me to study mimeographed copies of the then unpublished reports for the later years of the school.

[5]In 1789 it was reckoned that Rio had a population of about 40,000 and only four physicians. See Ernesto de Souza Campos, *op. cit.*, p. 17.

[6]For information on hospitals, see Pedro Sallos, *História da medicina no Brasil* (Belo Horizonte: Editôra G. Holman Ltda, 1971), pp. 85–122, and Lourival Ribeiro, *Medicina no Brasil colonial* (Rio de Janeiro: Editorial Sul Americana, 1971), pp. 33–86.

[7]Joseph F. X. Sigaud, *Du climat et des maladies du Brésil; ou Statistique médicale de cet empire* (Paris: Fortin, Masson, 1844), p. 4.

[8]Fernando Magalhães, *op. cit.* p. 102.

[9]Francisco Bruno Lobo (*ed.*), *Membria histórica dos accontecimentos mais notáveis occorridos na Faculdade de Medicina do Rio de Janeiro em 1879*, redigida pelo Dr. Nuno de Andrade, lente substituto da secção médica. Mimeographed copy, p. 2.

[10]An interesting evaluation of the physician's social status in this period is given by an English doctor in Bahia in 1852. He wrote:

> ". . . the Brazilian is a much more independent agent than the British physician. The higher orders too of the medical profession occupy a much more prominent position, *quoad* the public, than their brethren of Europe; and this, perhaps, may be accounted for by their superior education and knowledge of the world, as compared with the generality of the upper classes in a country situated like Brazil. The Brazilian physician, especially of the old school is generally well acquainted with European medical literature, especially the French. He is characterized by great liberality of feeling, he is little disposed to jealousy and altogether devoid of professional intrigue . . . In addition to his private practice, he commonly holds some independent public appointment or professorship; and, through the *concours* he seeks to rise by the approbation and respect of the profession, rather than by those humiliating practices by which, *on dit*, his European brother too often mounts to notoriety and fortune."

Robert Dundas, *Sketches of Brazil, including New Views on Tropical and European Fever, with Remarks on a Premature Decay of the System Incident to Europeans on their Return from Hot Climates* (London: John Churchill, 1852).

[11]According to a source cited by Guerra, the official or royal press, which was founded in 1808 by Dom João, and flourished under several different names in the colonial period (1808–1821), published 1,154 works in the period, of which only 48 were related in some way to the medical sciences. See Francisco Guerra, *Bibliografia médica brasileira: período colonial, 1808–1821* (New Haven, Connecticut: Yale University School of Medicine, 1958), p. 6. A large number of books written by Brazilian physicians were published in France, however, and after 1821 the volume published in Brazil increased. This medical literature badly needs study. For some of the works of medical writers, see Quarto Congresso Médico Latino Americano, *A medicina no Brasil* (Rio de Janeiro: Imprensa Nacional, 1908), pp. 57–74. See also Alvaro A. de Sousa Reis, *História da literatura médica brasileira* (Rio de Janeiro: Livraria J. Leite, n.d.). An outstanding contributor to clinical medicine was João Vicente Torres Homem. Among many other works, see his *Estudo clínico sôbre as febres do Rio de Janeiro* (Rio de Janeiro: Lopes do Couto e Cie, Editores, 1886).

[12]For a history of the Society of Medicine, later the Imperial and then the National Academy of Medicine, see Alfredo Nascimento, *op. cit.*

[13]Letter from Isabel Burton to her mother from São Paulo, May 3rd, 1868, published in W. H. Wilkins, *The Romance of Isabel, Lady Burton*, 2 vols. (New York: Dodd Mead and Co., 1897), Vol. 2, pp. 344–345.

[14]Isabel Burton, *The Life of Captain Sir Richard F. Burton*, 2 vols. (London: Chapman and Hall, Ltd., 1893), Vol. 1, p. 450.

[15]The complex history of public health legislation is another subject much requiring study. The massive work published under the sponsorship of Oswaldo Cruz when director of public health is invaluable; see Brazil, Directoria Geral de Saúde Pública, Placido Barbosa e Cassio Barboso de Rezende, *Os serviços de saúde pública, especialmente na cidade do Rio de Janeiro de 1808 à 1907 (Esboço histórico e legislação)*, 2 vols. (Rio de Janeiro: Imprensa Nacional, 1909).

[16]Joseph F. X. Sigaud, *op. cit.*, pp. 170–179.

[17]Stanley J. Stein, *Vassouras. A Brazilian Coffee County, 1850–1890* (New York: Atheneum, 1970), pp. 183–195.

[18]See Richard H. Shryock, *Medicine in America, Historical Essays* (Baltimore, Maryland: The Johns Hopkins Press, 1966), Chapter 3.

[19]See Charles E. Rosenberg, *The Cholera Years, The United States in 1832, 1849 and 1866* (Chicago: University of Chicago Press, 1962), for a fascinating account of the impact of cholera on the public health movement in the United States. For England, see Royston Lambert, *Sir John Simon, 1816–1904, and English Social Administration* (London: MacGibbon and Kee, 1963).

[20]Antônio Caldas Coni, *A escola tropicalista bahiana: Paterson, Wucherer, Silva Lima* (Bahia: Tipografia Beneditina Ltda, 1952).

[21]The hospital, Santa Casa de Misericordia in Rio de Janeiro, was at times a home for the medical school. Clinical teaching occurred in its wards.

[22]Quoted in Fernando Magalhães, *op. cit.*, p. 83.

[23]Francisco Bruno Lobo (*ed.*), *Memoria histórica . . . 1879*, reference 9.

[24]In 1868, Louis Pasteur had castigated the French government for failing to subsidize laboratory science. He wrote: "Laboratoires et découvertes sont des termes corrélatifs. Supprimez les laboratoires, les sciences physiques deviendront l'image de la stérilité et de la mort. Elles ne seront plus que des sciences d'enseignement, limitées et impuissantes et non des sciences de progrès et d'avenir. Rendez-leurs les laboratoires, et avec eux reparaîtra la vie, sa fécondité et sa puissance.

Hors de leurs laboratoires, le physicien et le chimiste sont des soldats sans armes sur le champ de bataille." See his "Le budget de la science," *Revue des cours scientifiques* 5 (1867–1868), 137.

[25]*Relatório do Ministro da Justiça e Negocios Interiores, J. J. Seabra, 1903-1904* in Primitivo Moacyr, *A instrução e a República*, 4 vols. (Rio de Janeiro: Imprensa Nacional, 1941–42), Vol. 3, *Código Epitácio Pessoa, 1900–1910*, p. 201.

[26]Abraham Flexner, *Medical Education in the United States and Canada; Report to the Carnegie Foundation for the Advancement of Teaching* (New York: Carnegie Foundation for the Advancement of Teaching, Bulletin No. 4, 1910).

[27]Louis Couty, "O ensino superior no Brasil," *Gazeta Médica da Bahia* 15 (Maio 1884), 521–532.

[28]Richard M. Morse, *From Community to Metropolis: A Biography of São Paulo, Brazil* (Gainesville, Florida: University of Florida Press, 1958), p. 162.

[29]The basic history of positivism in Brazil is found in João Cruz Costa, *Contribuição à história das idéias no Brasil (O desenvolvimento da filosofia no Brasil e a evolução histórica nacional)* (Rio de Janeiro: José Olympio, 1956), and his *O positivismo na República: notas sôbre a história do positivismo no Brasil* (São Paulo: Companhia Editôra Nacional, Biblioteca pedagógica brasileira, Serie 5a, Brasiliana 291, 1958). See also Ivan Lins, *História do positivismo no Brasil* (São Paulo: Companhia Editôra Nacional, Brasiliana 322, 1964). Auguste Comte, the chief exponent of positivist philosophy, believed that true knowledge was based on the methods and discoveries of the "positive" or physical sciences. Religious and metaphysical ideas were therefore not a form of knowledge. Progress would come through the study of science, which would form the foundation of the modern state. The motto of the Republic of Brazil in 1889 was "Order and Progress," after the positivist philosophy. The direct impact of positivism on science in Brazil, however, appears to have been slight.

[30]For a political history of the early years of the Republic, see José Maria Bello, *A History of Modern Brazil, 1889-1964* (Stanford, California: Stanford University Press, 1966), pp. 46-195, and João Pandiá Calógeras, *A History of Brazil* (New York: Russell and Russell, Inc., 1963), pp. 275-312.

[31]For a discussion of some aspects of racial pessimism in Brazil, see Thomas E. Skidmore, "Brazil's Search for Identity in the Old Republic" in Raymond S. Sayers (*ed.*), *Portugal and Brazil in Transition* (Minneapolis: University of Minnesota Press, 1968), p. 133.

[32]Quoted in George W. Stocking, Jr., *Race, Culture and Evolution: Essays in the History of Anthropology* (New York: The Free Press, 1968), p. 50.

[33]See Chapter 7 of this book.

[34]São Paulo, Secretário de Saúde e Assistência Social, *Correspondência de Emílio Ribas* (setembro 1946). Letter from Pereira Barreto dated 10/5/1900. At the time of writing, Barreto believed that water was a key factor in the transmission of yellow fever.

[35]Donald Cooper, *op. cit.*, 49.

[36]See A. Hunter Dupree, *Science in the Federal Government: A History of Policies and Activities to 1940* (Cambridge, Massachusetts: The Belknap Press of Harvard University Press, 1957), pp. 258-263, for the failure of the National Board of Health to take action over yellow fever, and the role of the American Public Health Association in the United States.

[37]Many South Americans claimed to have discovered the "bacillus" of yellow fever in the years after 1880. Eventually the North American bacteriologist George M. Sternberg was sent by the United States government in 1887, 1888, and 1889 to evaluate these claims. Sternberg's *Report on the Etiology and Prevention of Yellow Fever* was a masterly, detailed, but negative account of all the previous bacteriological work on the causes of the disease. Among those whose work he examined were the Brazilian Dr. Domingos Freire, and the Cuban physician Carlos Finlay, who in fact originated the hypothesis of the transmission of yellow fever by the *Aedes* mosquito. Both Freire and Finlay believed

they had isolated a yellow fever bacillus. In commenting on their work, Sternberg characterized some of the difficulties these early bacteriologists experienced in carrying out experimental work in South America:

> Like many pioneers in bacteriological research far from centers where modern exact methods had their origin . . . [Finlay] was not familiar with the methods of isolating and differentiating micro-organisms and fell into the usual and almost inevitable error of inference as to the source and genetic relation of various microorganisms.

See United States Treasury Department, Marine Hospital Service, *Report on the Etiology and Prevention of Yellow Fever* by George M. Sternberg (Washington: Government Printing Office, 1890), p. 32. It should be pointed out that Sternberg also believed for a time that he had discovered the "bacillus" of yellow fever.

[38]The announcement of the confirmation of the Finlay hypothesis was made by Walter Reed, *et al.* in, "The Etiology of Yellow Fever: A Preliminary Note," *American Public Health Association, Public Health Papers and Reports* 22 (1900), 37–53.

4

Epidemic Disease and the Growth of Science: The Serum Therapy Institute of Rio de Janeiro

It was not, after all, yellow fever that initiated the first steps in what, retrospectively, we see was the opening of a new era in institutional science in Brazil. It was, however, another crisis in public health, a crisis of sufficient proportions to threaten the economic interests of the elites, and to force public authorities to establish new institutions of medical science to cope with the situation. The solving of the enigma of the transmission of yellow fever by the Reed Commission only a few months later provided the context in which one of these new institutions, the Serum Therapy Institute of Rio, was able to thrive and carve out a new path for research.

In late 1899 rumors reached the capital city of Rio de Janeiro that bubonic plague had arrived in Santos, a city some three hundred miles to the southwest. Santos was a major entry point for European immigrants, many of whom had come to Brazil to work on coffee plantations in the interior of the state of São Paulo. The news of a possible epidemic of the unfamiliar and terrifying bubonic plague, at a time when immigration was vitally important to the booming coffee export industry, was a cause of grave concern. Politically and economically much was at stake.

The appearance of plague in Brazil signaled the arrival in America of the world pandemic of the disease that had started in the Far East in 1894. That year some one hundred thousand people died in Hong Kong, and as the disease spread to India another one million three hundred thousand died in the first two years of the epidemic.[1] The eruption of plague raised afresh the question of its cause and its mode of transmission. A series of medical commissions were sent into epidemic areas to study the disease, armed with the new techniques of bacteriology. In Hong Kong, the French bacteriologist Alexandre Yersin and the Japanese scientist S. Kitasato independently identified the plague bacillus (*Pasteurella pestis*) for the first time in 1894. Their discovery did not immediately result in the elimination of the disease, as plague is transmitted by a complicated process involving the flea, the rat, and human beings. In fact, ten years passed between Yersin and Kitasato's isolation of the plague bacillus and the final working out of the plague cycle. In that interval, however, advances in vaccination and serum therapy provided scientists with new insights into treatment. By 1896, a plague vaccine had been developed by Haffkine, and by 1898 Yersin employed the first anti-plague serums. The theory that the rat-flea transmits the plague bacillus to human beings from infected rats was first stated by Paul Louis Simond in 1898, though the actual stages of this cycle were only fully demonstrated in 1914. By 1900, control of bubonic was possible through the use of vaccines, serums, and the elimination of the rat.[2]

Few physicans in Brazil were familiar with the recent work on the plague. It is to the credit of the federal and local authorities that when action was required, the handful of physicians knowledgeable in the techniques of bacteriology were called into service. At the time of the arrival of the plague the federal and municipal public health services in Rio de Janeiro were not especially well prepared to deal with an epidemic. The state of São Paulo, however, had established a relatively comprehensive public health system in 1892, and was therefore in a somewhat better position to deal with outbreak of the plague.[3] News about the epidemic in Portugal and the rest of Europe led to quarantines in various of the ports to which European ships called in the months preceding the actual arrival of cases in Brazil.[4] It was, in fact, lucky that the first cases appeared in the state of São Paulo, rather than elsewhere, and the epidemic was checked by prompt official action. From São Paulo, the director of the Bacteriological Institute, Dr. Adolfo Lutz, was authorized to go to Santos with his student, Dr. Vital Brasil, in order to undertake studies of plague victims.[5] The federal government offered to help the state of São Paulo in its investigations. This offer was refused by the governor of São Paulo. When the federal authorities decided nonetheless to send the well-known Rio surgeon, Dr. Eduardo Chapot-Prévost, to Santos the decision was interpreted by Paulistas as an unwarranted interference in state matters.[6] In an editorial, the leading newspaper *O Estado de São Paulo* expressed the opinion,

> For us, the opinion of Dr. Chapot-Prévost cannot have the importance it has for those who charged him with the task of certifying

> whether or not our Bacteriological Institute is right or wrong. For us it is impossible to doubt that Dr. Adolpho Lutz and his intelligent and dedicated auxiliaries were right, and even if Dr. Chapot-Prévost were to assert that they had erred (which seems to us impossible), we would not lose any confidence in our functionaries.[7]

The confidence of the public in Lutz and the public health services was fully justified, though the spirit in which officials from the state of São Paulo responded to the efforts of the federal government to develop a national plan in public health was an important factor in reducing the efficiency of public health administration in Brazil in general. As a result of the clear diagnosis of plague in Santos, the São Paulo state authorities requisitioned a farm called Butantã on the outskirts of the city and appointed Lutz' student, Vital Brasil, in charge of organizing a small laboratory there, where vaccines and serums against the plague could be produced.[8]

Meanwhile plans for the defense of the city of Rio de Janeiro against the plague were in a state of confusion. According to the existing legislation, the federal government was responsible for the defense of the federal district and for carrying out quarantines in the ports of the Union. The municipality of Rio de Janeiro had responsibility for the sanitation of the city itself, and the federal government could only intervene in the city at the request of the city authorities. Inadequate funding and staffing of the federal public hygiene department, and the separation of the different tasks of sanitary defense and attack, made an efficient solution to the crisis caused by the eruption of the plague extremely difficult.[9] Both the federal and municipal authorities took it upon themselves to send medical commissions to Santos to study the situation. The work of Dr. Chapot-Prévost has already been mentioned.[10] The municipality commissioned a young bacteriologist just returned from almost three years' study abroad, Dr. Oswaldo Gonçalves Cruz, to accompany Chapot-Prévost to Santos.[11] In Santos, the two doctors installed themselves in the Isolation Hospital and began autopsies of plague victims. Shortly afterwards they confirmed the findings already made by Lutz that plague had indeed arrived in Brazil.[12] The first cases of plague in the urban zone of Rio de Janeiro occurred in January and February of 1900, and two hundred and ninety-five people died that year in the capital.[13]

The federal government was immediately requested by the

Prefecture in Rio de Janeiro to provide funds to protect the city of over six hundred thousand people. The intervention of the federal government in municipal affairs was shortly afterwards suspended, according to the semi-official account of the situation made while Cruz was director of public health a few years later, but a special sum of money was contributed to the municipality for the specific purpose of dealing with the plague. The "dual and anarchic" organization of the federal and municipal hygiene services in Rio in fact rendered efficient sanitary action against epidemic disease virtually impossible, until the services were united and organized under Cruz in 1903.[14]

Meanwhile it was decided by the municipality to establish a small laboratory where vaccines and serums against plague could be produced at small cost. On the advice of Baron Pedro Affonso, director of the Instituto Vaccínico do Distrito Federal (Federal Vaccination Institute), a farm some distance from the city was requisitioned.[15] The farm, known as Manguinhos, had originally been acquired by the municipality as a site for an incinerator, which had never been completed. On the hilly and overgrown land there stood a few tumbledown buildings to serve as laboratories where work on the dangerous plague bacillus could begin at a safe distance from the center of town.

No sooner had this important project been decided upon, however, when a new epidemic forced the municipality to turn its attention and financial resources elsewhere. An outbreak of cattle disease among the herds in the state of Rio de Janeiro led officials to apply again to the federal government and suggest it take over the installation of the laboratory. On May 24, 1900, the proposal was accepted, and Baron Pedro Affonso was formally authorized by the federal authorities to continue as director of the project and recruit a staff and equip laboratories. The Instituto Sorôtherápico Federal de Manguinhos (Federal Serum Therapy Institute of Manguinhos) was given its official name and attached to the federal department of public health as a dependency of the Ministry of Justice and the Interior.[16]

To recruit a trained bacteriologist as chief of the technical staff of the laboratory, the Baron first looked to France, and specifically to the Pasteur Institute in Paris. In this task he was not successful. As he explained in a note to the Minister of the Interior, no French bacteriologist of repute was willing to come to Brazil on the six-month contract offered by the government, to an area where the risk of yellow fever was still great.[17] At this moment, however, Professor Emile Roux, Pasteur's distinguished pupil and vice-director of the Pasteur Institute, recalled that the Brazilian doctor Oswaldo Cruz had just returned to Brazil after completing training at the Pasteur Institute, and suggested his name to the Baron. Cruz was duly offered the post of chief bacteriologist at Manguinhos, which he accepted after Pedro Affonso agreed to Cruz' list of necessary equipment.[18] By May 1900, at the age of twenty-eight, Cruz therefore found himself launched on a career as a microbiologist at a "crisis" institution with the limited and practical goal of preparing vaccines and serums for supply to the federal government at low cost. There was no guarantee that the institute would survive the containment of the plague epidemic. In fact, from this moment on Cruz' life was almost entirely absorbed by the work and future development of the institute.

Oswaldo Cruz and His Training

Oswaldo Gonçalves Cruz was born on August 5th, 1872, in the state of São Paulo, in the small town of São Luís de Paraitinga, where his father, Dr. Bento Gonçalves Cruz, was in private medical practice.[19] In 1877, the family had moved back to the father's home state of Rio de Janeiro. Here it was that Cruz grew up. His father continued in private practice except for a period in the field of public health as a member of the Junta Central de Higiene Pública. By 1890, he had been appointed as assistant to the Inspector General of Hygiene, and in 1892, a few months before his death, was named Inspector General of Hygiene. Given his father's profession and the tendency of sons to follow in the same profession in Brazil, it was perhaps not surprising that Cruz entered the Rio Medical School at the age of seventeen. Here he came into contact with a number of the well-known clinicians. Perhaps guided by his father's interest in the sanitation sciences, Cruz also became a *préparateur* in the Hygiene Laboratory attached to the chair of Hygiene in the school, the chair being occupied by Professor Rocha Faria. As part of his examination for graduation, he prepared his medical thesis under Rocha Faria on the subject of the transmission of microbes by water.[20]

By 1892, the year of his graduation, Cruz had exhausted the training facilities in Brazil. He was a quiet, thoughtful young man, distinguished from his fellow-students mainly by his serious commitment to experimental medicine.[21] Any hopes he may have entertained of going abroad for further training were dashed, however, by his father's death that same year. At twenty years of age Cruz found himself already married (he married Emília Fonseca in 1893) and responsible for his family. He turned to clinical medicine with a sense of obligation to his own family and his father, following his interest in microbiology only in his spare time in a small laboratory his maternal grandfather had equipped for him in his home as a wedding gift. In this way he worked for the next four years. Between 1892, the year of his graduation, and 1899 when he was contacted by the new serum therapy institute, he published fourteen scientific articles, many of them of a practical and descriptive nature. He also became a member of an informal group of physicians who studied German at the Rio Polyclinic (Policlínica Geral de Rio de Janeiro). There his attempts to undertake laboratory analysis as an aid in diagnosis were encouraged by his friend, Sales Guerra.[22]

Cruz' life was changed dramatically in 1896, when his grandfather offered to send Cruz and his family to Paris so that Cruz could obtain specialized training in microbiology at the Pasteur Institute, one of the most famous centers of microbiological research. Cruz' move to Paris for training was, of course, in the tradition of the colonial scientist for whom the centers of the scientific world still lay outside his own country. In deciding to go to Paris, Cruz was encouraged by his former professor, Dr. Francisco de Castro. His return two and a half years later as a fully trained bacteriologist and nationalist dedicated to the advance of science within Brazil was to make his stay in Paris of great significance.

Several factors may have brought Cruz to France rather than Germany, where so many of the young scientists from the United States were going for advanced medical training.[23] First were the many cultural ties between Brazil and France, dating from the beginning of the nineteenth century. Brazilians had always looked to France for intellectual and cultural leadership, and French was the educated Brazilian's second language. Second was the special tie between Brazil and Pasteur. In 1883, the Emperor Dom Pedro II had invited Pasteur to come to Brazil to study the yellow fever that regularly devastated the population of Rio de Janeiro. Pasteur declined the invitation on the grounds of old age and ill health, but a young Brazilian doctor, Augusto Ferreira dos Santos, was sent by Dom Pedro in 1886 to study

Pasteur's rabies' inoculation methods.[24] On his return, Ferreira dos Santos inaugurated a Pasteur Institute in Rio for the treatment of hydrophobia. When the Pasteur Institute of Paris was founded by public subscription in honor of Pasteur, Dom Pedro was among those commemorating his admiration for the French scientist by a contribution. When Cruz arrived at the Pasteur Institute in the winter of 1896, therefore, his nationality assured him a warm welcome. In Paris he lived for the next two and a half years, attending courses in bacteriology and hygiene sciences at the Pasteur Institute and at the Paris Municipal Laboratory.

In 1896 the reputation of the Pasteur Institute as a major center of research was already well-established. The Institute had originated in 1885 when Pasteur had announced he had found a cure for rabies.[25] The Académie des Sciences was informed by Pasteur that of the three hundred and fifty victims already treated for the bite of mad dogs, only one had died. Pasteur suggested to the Academy that a model institute might be founded for the treatment of hydrophobia. A committee, which included Vulpian and Charcot among its members, was assembled to discuss the project. It had already been agreed that an institute be organized bearing Pasteur's name when Pasteur suffered a slight stroke in 1887. The public response to the news of Pasteur's illness was immediate. The institute, which was funded entirely through private donations of money, received such large sums that by 1888 the buildings were ready for occupation. Emile Roux and Charles Chamberland, two of Pasteur's most brilliant disciples, and the Russian bacteriologist Elie Metchnikoff, were among those named to the staff, with Pasteur as director. It was decided that the institute should have the right to sell all the vaccines it produced for a profit, in order to finance scientific research. The institute also began to publish its own journal, the *Annales de l'Institut Pasteur*, as it was felt that the official journal of the Académie des Sciences was too formal a vehicle for reporting research news from the institute.[26]

Originally, the work of the Pasteur Institute was divided among six departments—the study of rabies under Grancher, general microbiology under Duclaux, who also acted as Vice-Director, the techniques of microbiology under Roux, the applications of microbiology under Chamberland, the morphology of microbes under Metchnikoff, and comparative microbiology under Gamaleia (the last department was in fact never organized). Theoretically, therefore, there was a permanent research staff of six, excluding Pasteur. The basic organizing idea of the Pasteur Institute was to combine original research, the application of research to problems of practical hygiene (via the production of new vaccines, for instance), and the training of students, in a single institution. Each staff member was expected to work in his own area of research while cooperating as fully as possible with other researchers.

After many years of ill health, Pasteur died in 1895, the year before Cruz reached the institute. The direction of the institute was officially taken over by Duclaux. After 1888 the search for new protective serums intensified; as in an earlier period, when the search for new pathogenic organisms dominated bacteriology, the rapid development of serum therapy raised some false hopes and resulted in the development of useless serums. In the long run, however, the focus on serum therapy was fruitful, and led to the development of anti-tetanus, anti-diptheria, anti-streptococcus, and anti-plague serums, and opened up the field of immunology. Cruz arrived at the Pasteur Institute when interest in serum therapy was at its height, and his training in the most advanced techniques in this branch of medicine prepared the way for his appointment to the Serum Therapy Institute at Manguinhos in 1900.

Another distinctive feature of the Pasteur Institute was its courses of instruction. Perhaps the most famous was the "cours de microbie technique" taught by Emile Roux, initiated in 1889 as a series of thirty lectures repeated three times each year. As the students increased in number and as knowledge grew, the numbers of lectures also grew, and from 1894 on the course was offered by Roux once a year to as many as one hundred students at a time. The lectures were followed by laboratory demonstrations and practical work supervised by Roux, and aided by *préparateurs* such as Yersin. The close association between researchers and students brought the students into direct contact with some of the most original scientific minds of the day.

It appears that Cruz thrived at the Pasteur Institute. Yet there seems to have been little doubt of Cruz' intention to return to Brazil at the end of his allotted time. From Paris, he had written to Sales Guerra that he was studying hygiene, microbiology, pathology, histology, and chemistry, with the intention of mounting a laboratory for the diagnosis of morbid diseases on his return to Rio de Janeiro.[27] Another constant preoccupation drawing him back to Brazil was yellow fever. "When will we free ourselves from this plague?" he wrote. "It is our 'Nessus coat.' It is like an indelible stain that disgraces and humiliates us."[28] Yellow fever was to be the thread that linked so many of the activities that filled his life. Yet in planning to return to Brazil, Cruz must have been under few illusions about the opportunities for a career in research. While in Paris, he attended lectures in urinology by Professor Guyon, though as he remarked to Sales Guerra, he "abominated clinical medicine."[29] As if foreseeing a future need, he also donned the workingman's blue shirt and visited a glassblowing factory to learn how to prepare his own equipment for bacteriological research.[30]

Early Years at the Serum Therapy Institute

Cruz' return to Brazil in the fall of 1899 as a trained, highly motivated microbiologist committed to the belief that science should be an integral part of national culture was a factor of great importance for the growth of science in Brazil. The status and prospects of research in Brazil, however, had not improved during his absence. It must have been with some reluctance that he settled into clinical work, establishing, as he had announced, a small laboratory where he undertook pathological and bacteriological examinations for physicians in the area. It was not until the late spring of 1900, when he was offered the position at the newly founded Serum Therapy Institute of Manguinhos, that Cruz' career as a full-time bacteriologist was launched.

When Oswaldo Cruz joined the laboratory and was placed in charge of the technical production of vaccines and serums, the staff consisted of himself, an army bacteriologist called Dr. Ismael da Rocha, and two young medical students, Henrique de Figueiredo Vasconcellos, who had come from the Vaccination Institute run by the Baron, and Ezequiel Caetano Dias. A short while after Cruz arrived, a fresh outbreak of plague among the armed forces resulted in the recall of Dr. Rocha, and his place was taken by a third medical student, Antônio Cardoso Fontes.

To reach the laboratory at Manguinhos the staff had to make a journey of about eight miles by train, followed by a horseback ride or walk to the hilly estate where their equipment was installed. The total annual budget was sixty-thousand Brazilian *milreis*, or about fifteen thousand dollars at the rate of exchange then current.[31] Horses were acquired for the preparation of serums, and the work of the institute begun. By October of the first year, only six months after the institute's inauguration, the first boxes of vaccines prepared by Cruz were delivered by Baron Pedro Affonso to the director of public health in the capital.[32] By February the following year, anti-plague serums were also being supplied. The techniques used in their production were described in the Rio medical journal, *Brasil-Médico* that year, as the first official publication of the laboratory.[33] By the end of 1903, according to a note prepared by Cruz for his own reference, materials worth an estimated 324,179 milreis (or just over eighty thousand dollars) had been sent at no cost to the government in the space of just over two years.[34]

Some fatal accidents involving anti-plague serums prepared at Manguinhos occurred in the first months that might have undermined Cruz' reputation as a competent bacteriologist, had he not quickly taken steps to investigate the matter. In one of the cases, involving the death of a well-known Rio surgeon, Cruz was able to establish that death was caused not by any toxicity in the serum itself, but by an incompatibility between the victim's blood and the serum itself. His findings and his discussion of deaths in cases involving serums in Europe and Brazil were published in 1902 in the journal *Brasil-Médico* and seem to have allayed alarm.[35] At the time nothing was known about blood types and serums were injected in massive doses straight into the vein.

Within the laboratory Cruz quickly began to exert a strong influence on the staff. At the age of twenty-nine, as he was in 1901, Cruz was a stocky man of medium height, with an arresting and rather fleshy face. He seemed older than his years, and was usually reserved and somewhat silent. His black hair was already greying and he wore it rather long and flowing. His elegant European-style black coat, so unsuitable for the Brazilian heat yet fashionable among the Brazilian upper classes, was offset by a soft white cravat tied in a loose knot at the neck. For all his reserve, he was a compelling figure, a man of intensely felt enthusiasm for science, of great resourcefulness, able to communicate this enthusiasm and the need for hard work and accuracy to others. He was a natural teacher and had a lifelong influence on those who worked closely with him.

Cruz' personal and scientific authority soon led to difficulties between himself and the Director, Baron Pedro Affonso. The Baron was a competent medical doctor, but an irascible administrator with only a small knowledge of the more advanced techniques of serum therapy. A few minor clashes of temperament and will between the two men led to the Baron's resignation at the end of 1902, since Cruz' departure threatened the entire enterprise.[36] Cruz was formally appointed director in early 1903.

Models in Science: The Pasteur Institute

No sooner had Cruz been appointed director of the Serum Therapy Institute of Manguinhos than he expressed dissatisfaction with the institute's limited function as a supplier of vaccines and serums. From his training at the Pasteur Institute in Paris Cruz had acquired a knowledge of the organization of one of the most successful scientific institutions in the world. Ben-David has pointed out that the Pasteur Institute was one of the first of a new kind of research institution, in which the research carried out was not basic, in the then accepted sense of the term, but instead was "applied" or problem-oriented.[37] While the emphasis still lay in original discovery, the focus lay on practical health problems. As an example of the new problem-oriented institution of scientific research the Pasteur Institute had an obvious appeal to developing countries. The combining of pure and applied research with student training, its freedom from bureaucratic restrictions, and its financing for research also made it seem an ideal "model" for a similar institution in Brazil. That the Pasteur Institute was much on Cruz' mind as a model for the development of the Serum Therapy Institute is shown by a note written by Cruz to the Brazilian Congress in late 1903, in which he proposed certain changes in the organization and purposes of the institute. Manguinhos, he argued, should be

> transformed into an institute for the study of infectious and tropical diseases, along the lines of the Pasteur Institute of Paris . . . The Institute should be charged with the preparation of all therapeutic serums, vaccines, with anti-rabies treatment, the preparation of industrial ferments, with the teaching of bacteriology and parasitology, and would transform itself into a nucleus of experimental studies that would greatly enhance the name of our country abroad.[38]

His experience at the Pasteur Institute, it is evident, had provided him not only with the skills to undertake such a project, but the desire to find the opportunity to put his skills into practice. The professionalization of the career of experimental medicine thus became an additional stimulus to the growth of science.

The relation between the Pasteur Institute and the evolution of the Oswaldo Cruz Institute in Brazil, however, needs further elaboration. Some of the most crucial questions about the institutionalization of science in developing countries concern the processes by which elements of successful institutions organized elsewhere can be incorporated into a totally different culture. At the same time, the historical rise of independent research institutes as one of the chief loci of scientific work raised the possibility of the establishment of autonomous research organizations at a time when the general appreciation of the value of science was still low. For both of these reasons, we must ask to what extent the growth of the Serum Therapy Institute into the "nucleus of experimental studies" envisaged by Cruz in 1903 represented merely a transfer of the model of the Pasteur Institute to Brazil.

In fact, the different social, political, and intellectual circumstances of the two countries make an analysis along these lines impossible. Furthermore, it would beg the entire question of why Cruz was able to succeed in his ambitions for science while similar institutions in Brazil being founded at this time were failing. First we have to consider the facts surrounding the founding of the Pasteur Institute in Paris. The Pasteur Institute was an anomaly in France, a privately funded and totally independent institution of scientific research. Its founding was the result of the enormous prestige which Pasteur enjoyed because of the influence his scientific work had on economic questions in France and on the universal problems of illness and death. By comparison, when Cruz requested the funding of a Pasteur Institute in Brazil in 1903 he was virtually unknown in the world of international science and relatively unknown in Brazil except in his small circle of scientific and medical friends. When Cruz' name was put forward as a possible candidate for the post of director of public health, few officials knew who he was.[39] Another important difference concerned the patronage of science. No tradition of private patronage of science existed in Brazil, and the value attached to science was too slight to make this a feasible method of encouraging the growth of science. No Carnegies or Rockefellers existed whose fortunes acquired in industry could now be put to the service of research science. Nor, as was the case with Pasteur, was Cruz sufficiently well known to start a new tradition of patronage of science by the elites.

In Brazil, in fact, the only likely sources of support for science were the national and state governments, and any future developments in science were likely therefore to occur in government scientific agencies. As a government activity, science would have to become a competitor with other agencies for scarce resources. This raised a host of questions concerning the political visibility of science, the role of the politician as a promoter of science, and the general legitimacy of science as a social endeavor. It also raised the question of whether a purely research-oriented institute would find support, rather than an institution, as the Serum Therapy Institute was in 1903, devoted to the practical application of techniques developed outside Brazil. The existence of an organization such as the Pasteur Institute, totally independent of government regulation and scrutiny by non-scientists, was also unlikely in Brazil.

In considering the relation between the "model" of the Pasteur Institute and the rise of the Oswaldo Cruz Institute in Brazil, a better case might perhaps be made for comparing the Brazilian institution with Pasteur Institutes being founded in this period in several countries by the French colonial government. Yet here, too, the parallel is misleading. The projection of the bacteriological and affiliated sciences into the world at large was an almost inevitable accompaniment of the development of the sciences themselves, because of their great consequences for the control and treatment of disease. The German bacteriologist Robert Koch, for instance, spent a total of seven years outside Germany between 1896 and 1907, studying the transmission of disease in such places as South Africa, Bombay, Dar-es-Salaam, Italy, and Rhodesia. The isolation of the cholera vibrio was one discovery to come out of his medical missions abroad. The Pasteur Institute itself sponsored several scientific missions, such as Yersin's to Hong Kong in 1894, from which came the discovery of the plague bacillus.[40] Schools of hygiene and tropical medicine were founded in several European countries, especially in those countries extending their colonial possessions in tropical areas.[41] With the scramble for Africa at the end of the century, and the extension of French colonial administration in the Far East, French colonial officials were led to consider such questions as the white man's adaptation to tropical climates and the effect of tropical disease. One result was the establishment of a number of permanent Pasteur Institutes in China, Vietnam, and Africa, all closely modeled on the original Institute in Paris and all with some official ties to it. The staffs of these institutions were usually recruited directly from the Pasteur Institute in Paris.[42]

Brazil possessed no such cadre of trained bacteriologists to aid in the organization of a Pasteur Institute in Rio de Janeiro in 1903. Nor was the Brazilian government interested in such a development. Congress' answer to Cruz' request in 1903 to make the Serum Therapy Institute into a Pasteur Institute for Brazil was a resounding "No."[43] Without the support of the government, even had a group of foreign bacteriologists come to Brazil to form a nucleus of experimental medicine, it is highly doubtful they would have been able to *create* the conditions for such government support. Foreign scientists are more likely to be effective in aiding the development of science in a developing country, that is, once a decision has been made by a national government to develop science as a matter of policy. Given the existing cultural obstacles to science in Brazil, and the absence of any plan to expand scientific activities on the part of the government, a Brazilian scientist able to recruit other Brazilian scientists and with a good grasp of the tasks of institution-building was perhaps in a better position to *initiate* a new phase in government policy toward science.[44]

Enough has probably been said to show that the mere existence of models of scientific institutions cannot guarantee their successful imitation and implantation in foreign cultures. Imitation of institutions has, of course, played an important role in the growth of science.[45] But unless other conditions within the country favor the development of science, mere imitation can result in stagnation, as occurred, for instance, with the National Museum. Successful imitation involves adaptation and transformation of models to new political and cultural realities. The process involves sensitivity to the traditional barriers to science, institutional entrepreneurship, political acumen, and administrative ability. In the case of the Serum Therapy Institute, any evolution into a center of experimental studies would have to be the result not merely of Cruz' desire to create a Pasteur Institute, but his ability to convince legislators of the usefulness of such an institution, secure financial support and a degree of administrative autonomy, recruit a trained staff, and institute mechanisms to ensure the continued recruitment of scientists and financial supports over time in order to prevent stagnation.

The history of the Serum Therapy Institute between 1900 and 1903 in fact falls into a classic mold. Like many institutions in the field of public health, it owed its existence to a crisis that seemed to offer resolution by the founding of a new institution of science. Here a crisis in public health might be contrasted with a crisis in other fields, such as education. A crisis in education may be just as profound, but its causes are likely to be so complex and its solution so perplexing as to tax resources and bewilder administrators and the public. Such a crisis is therefore unlikely to be resolved in a clear-cut fashion by the creation of a narrowly defined program with a definite practical goal. Because physicians form the bulk of the trained scientists in most developing countries, it is also not surprising that new scientific developments often occur first in the field of medicine.[46]

As medical science advanced, increasingly, medical problems too, were seen to be capable of solution. In a sense, the history of the Serum Therapy Institute and the Butantã Institute in São Paulo between 1900 and 1903 can be seen as an extension of the colonial tradition described by Basalla into the period of scientific and experimental medicine.[47] Both of these institutions were colonial in their reliance on techniques and traditions of science developed abroad. By 1903, neither institution possessed the capacity to carry out original research, or even applied science on a large scale.

In addition to being "colonial" institutions, both also displayed many of the characteristics of "crisis" institutions. While crises in public health often generate immediate support for new laboratories, these laboratories are usually given such restricted functions, budgets, and staff as to limit seriously any expansion once the crisis has passed. This is especially true when institutions use only practical methods developed elsewhere. Crisis institutions therefore run the risk of stagnation. To a certain extent, this is what happened with the Butantã Institute of São Paulo, the earlier Pasteur Institute for the treatment of rabies, and the Vaccination Institute for inoculation against smallpox.[48] The position of the Serum Therapy Institute in 1903 differed only because of Cruz' strategic foothold in a federal scientific establishment, and his ambition to see the Serum Therapy Institute grow. The rest of the story lay in the future.

References

[1]Charles Singer and E. Ashworth Underwood, *A Short History of Medicine* (New York and Oxford: Oxford University Press, Second Edition, 1962), pp. 488–492.

[2]*Ibid.*, pp. 411–420, has a short discussion of the development of theories of immunization and serum therapy. Serum therapy followed the discovery by Emil von Behring and Kitasato in Koch's laboratory that non-lethal doses of toxin injected into the serum of an animal would provide a serum giving protection from the disease. Eventually the horse was used to develop therapeutic sera.

[3]See Chapter 7.

[4]Brazil, Directoria Geral de Saúde Pública, Placido Barboso e Cassio Barboso de Rezende, *Os serviços de saúde pública no Brasil, especialmente na cidade do Rio de Janeiro de 1808 à 1907* (*Esboço histórico e legislação*), 2 vols. (Rio de Janeiro: Imprensa Nacional, 1909), Vol. 1, p. 104.

[5]An account of the work of Adolfo Lutz and Vital Brasil in Santos is found in São Paulo, Instituto Adolfo Lutz, Departmento do Administração, Vital Brasil, *Relatório sôbre a peste bubônica em Santos, apresentado ao Dr. Director do Instituto Bacteriológico, em 27 de novembro de 1899.* Handwritten original.

[6]The news of Chapot-Prévost's arrival, and the reaction of the São Paulo public to the rumor that Chapot-Prévost found the city of Santos in a panic, is described in the newspaper *O Estado de São Paulo*, October and November issues, 1899.

[7]*O Estado de São Paulo*, 5 November 1899.

[8]The subsequent history of the Butantã Institute is described in Chapter 7.

[9]Brazil, Directoria Geral de Saúde Pública, *Os serviços de saúde pública no Brasil*, Vol. 1, pp. 97–104.

[10]Eduardo Chapot-Prévost (1864–1907) had traveled abroad with the Brazilian bacteriologist Dr. Domingos Freire, and took a keen interest in the bacteriological sciences. Biographical details are found in Fernando Magalhães, *O centenário da Faculdade de Medicina do Rio de Janeiro, 1832–1932* (Rio de Janeiro: A. P. Barthel, 1932), pp. 227–228.

[11]See J. Guilherme Lacorte, "A atuação de Oswaldo Cruz no aparecimento da peste bubônica no Brasil," *A Folha Médica* 54 (Fevereiro 1967), 183–188.

[12]Cruz' official report made to the federal authorities, containing a detailed account of the steps taken to identify the plague, was published as Oswaldo Gonçalves Cruz, *Relatório acêrca da moléstia reinante em Santos, apresentado pelo Dr. Oswaldo Gonçalves Cruz à S. Ex. O Sr. Ministro da Justiça e Negocios Interiores* (Rio de Janeiro: Imprensa Nacional, 1900).

[13]Brazil, Directoria Geral de Saúde Pública, *Annuario de estatistica demographo-sanitária*, 1907, p. 95.

[14]Brazil, Directoria Geral de Saúde Pública, *Os serviços de saúde pública no Brasil*, Vol. 1, p. 106.

[15]Baron Pedro Affonso, as Director of the famous charity hospital in Rio, the Santa Casa de Misericordia, began to carry out inoculations against smallpox at the Vaccination Institute in 1887. From 1888 on work at the Vaccination Institute was supported by a federal stipend. According to his own report, a total of some 173,000 persons were vaccinated at the Institute between 1887–1908. Not all of these vaccinated, of course, remained in the city. See Barão Pedro Affonso, *Relatórios dos trabalhos do Instituto Vaccínico do Distrito Federal, segundo de um retrospetivo dos trabalhos vaccínicos de 1887 à 1917* (Rio de Janeiro: n. p. 1917), pp. 31–43.

[16]The history of the Oswaldo Cruz Institute, especially in its earliest years, has been drawn to a large extent from the *Oswaldo Cruz Files, Administrative Records* and *Museum Documents*. See the bibliography for a description of these manuscript sources. The reasons for the offer of the laboratory to the federal government are described in an Official Note from the Prefecture, 23 April 1900. Instituto Oswaldo Cruz, *Oswaldo Cruz Files, No. 7.* The terms of the exchange and the acceptance by the federal government are contained in an Official Note on May 9, 1900. *Oswaldo Cruz Files, No. 7.*

[17]Official Note from Director Pedro Affonso to the Federal Government, Instituto Oswaldo Cruz, *Oswaldo Cruz Files, No. 7.* March 2, 1900. A French veterinarian, Carrère, was brought to Brazil, but he found the climate of Rio de Janeiro unsuitable, and left shortly afterwards.

[18]E. Sales Guerra, *Osvaldo Cruz* (Rio de Janeiro: Case Editôra Vecchi Limitada, 1940), pp. 51–52.

[19]There are numerous biographical sources in Portuguese of varying quality and reliability for Oswaldo Cruz. The standard biography is by E. Sales Guerra, *op. cit.* Sales Guerra was a contemporary of Cruz, a close friend, and his book is highly informative. It does not, however, cite sources. A definitive biography in Portuguese or English, based on primary sources and thoroughly documented, would be a welcome addition to our knowledge of Brazilian history. Other useful sources are: Ezequiel Caetano Dias, *Traços biográficos de Oswaldo Cruz* (Rio de Janeiro: Imprensa Nacional, 1945); Gastão Perreira da Silva, *O romance de Oswaldo Cruz* (Rio de Janeiro: Brasília Editôra, n.d.); Antônio Austregesílio, *Oswaldo Cruz: vida gloriosa de Oswaldo Cruz* (Rio de Janeiro: Departmento Nacional de Saúde, 1937); Belisário Penna, *Oswaldo Cruz, impressões de um discípulo* (Rio de Janeiro: Revista dos Tribunaes, 1922); and Clementino Fraga, *Vida e obra de Osvaldo Cruz* (Rio de Janeiro: José Olympio, 1972). In addition, the centennial of Cruz' birth saw the republication of Cruz' scientific papers in Oswaldo Gonçalves Cruz, *Opera omnia* (Rio de Janeiro: Impressora Brasileira Ltda., 1972), and the reproduction, with commentaries, of many other documents relating to the sanitary movement in Rio and the Oswaldo Cruz Institute. See Edgard de Cerqueira Falcão, *Oswaldo Cruz; monumenta histórica*, 3 vols. (Sao Paulo: Brasiliensia Documenta 6, 1971–73).

[20]The thesis was entitled *Vehiculação microbiana pelas aguas*. These apresentada à Faculdade de Medicina do Rio de Janeiro, em 8 de novembro de 1892 (Rio de Janeiro: Papelaria e Impressora, 1893). It forms part of the collection of medical theses in the Rio Medical School library, and has recently been republished in Oswaldo Gonçalves Cruz, *Opera omnia*, pp. 19–199.

[21]By the time he had completed his thesis for graduation, Cruz had already published several small medical reports and reviews in Brazil. See Oswaldo Gonçalves Cruz, *Opera omnia*, p. 80.

[22]E. Sales Guerra, *op. cit.*, p. 29. The Policlínica Geral do Rio de Janeiro was founded in 1881 and offered free medical care to the poor.

[23]For a discussion of the role of Germany in the development of new institutions of science in America from 1870 on, see Thomas N. Bonner, *American Doctors and German Universities, A Chapter in International Relations, 1870–1914* (Lincoln: University of Nebraska Press, 1963). Between 1870 and 1914 no less than 15,000 Americans studied in Germany. Although the majority went to study clinical medicine, the influence of those studying experimental medicine far outweighed their numerical importance. In particular, at the Johns Hopkins Medical School the German influence was strong—of the new faculty appointed there in 1893, Kelly, Mall, Abel, Halstead, Welch, and Osler had all been to Germany.

[24]Some of this correspondence is reprinted in Georges Raeders, *Pedro II e os sábios franceses* (Rio de Janeiro: Atlântica Editôra, 1944) Anexo I. According to Dr. João Batista Lacerda, Director of the National Museum between 1892 and 1915, in the letter in which Pasteur excused himself from the invitation, Pasteur asked Dom Pedro II whether it would be difficult to make experiments on human beings using condemned Brazilian criminals in prison. Dom Pedro supposedly remarked that he viewed such experiments as a crime against the sacred right of human life. João Batista de Lacerda, *Fastos do Museu Nacional do Rio de Janeiro: recordações históricas e scientíficas fundadas em documentos authênticos e informações verídicas*. Obra executada por indicação e sob o patronato do Sr. Ministro do Interior Dr. J. J. Seabra (Rio de Janeiro: Imprensa Nacional, 1905), p. 129.

[25]The history of the founding and organization of the Pasteur Institute is taken from Albert Delaunay, *L'Institut Pasteur, des origines à aujourd'hui* (Paris: Editions France-Empire, 1962), and A. Calmette, "Institut Pasteur" in France, Ministère du Travail, *Livre d'or de la commémoration nationale du centenaire de la naissance de Pasteur* (Paris: Imprimerie Nationale, 1928).

[26]In 1903 a second publication was established called the *Bulletin de l'Institut Pasteur.*

[27]Quoted in E. Sales Guerra, *op. cit.*, p. 33.

[28]*Ibid.*, p. 33. He referred in the same letter to the investigations of the Italian Sanarelli into the "bacillus" of yellow fever.

[29]*Ibid.*, p. 31.

[30]*Ibid.*, p. 47. At a much later date, one of the technicians trained in glassmaking at the Institute by Oswaldo Cruz retired from the Institute to start one of the first general commercial agencies for supplying glass materials to the medical faculties and other scientific institutions in Brazil. Dr. Oswaldo Cruz Filho, in conversation to the author. See also E. Sales Guerra, *op. cit.*, p. 47.

[31]The monetary unit at the time was the *milreis*, written 1$000. Budgetary figures were often expressed in terms of 1,000 *milreis*, called the *conto*, written 1,000$000. According to Richard M. Morse, *From Community to Metropolis: A Biography of São Paulo, Brazil* (Gainesville: University of Florida Press, 1958), p. vii, the value of the *milreis* during much of the nineteenth century remained near 50 American cents. In the 1890s its value declined by as much as three-fourths; by 1905 it was worth approximately 25 cents, but its value increased somewhat before World War I. The calculation of the

institute's annual budget made here is based on a value of 0.25 dollars per *milreis.*

[32]Official Note, October 30, 1900. Instituto Oswaldo Cruz, *Oswaldo Cruz Files, No. 7.*

[33]Oswaldo Gonçalves Cruz, *A vaccinação anti-pestosa. Trabalho do Instituto Sôrotherápico Federal do Rio de Janeiro* (*Instituto de Manguinhos*) (Rio de Janeiro: Besnard Frères, 1901). Already the name Instituto de Manguinhos was being used interchangeably with the name Instituto Sôrotherápico Federal.

[34]Instituto Oswaldo Cruz, *Oswaldo Cruz Files, No. 7a,* Manuscript by Oswaldo Cruz. No date.

[35]The accidents were reported as *Dos accidentes em sôrotherápia. Trabalho do Instituto Sôrotherápico Federal do Rio de Janeiro* (Rio de Janeiro: Besnard Frères, 1902).

[36]Pedro Affonso's official departure occurred in December 9, 1902. Instituto Oswaldo Cruz, *Oswaldo Cruz Files, No. 7.*

[37]Joseph Ben-David, *The Scientist's Role in Society. A Comparative Study* (Englewood Cliffs, New Jersey: Prentice-Hall, Inc., Foundations of Modern Sociology Series, 1971), pp. 142–143.

[38]Brazil, Congresso Nacional, *Annaes da Câmara dos Deputados,* Vol. 5, Sessões de 1 à 30 de setembro de 1903, p. 588.

[39]E. Sales Guerra, *op. cit.,* p. 62.

[40]The Pasteur Institute sent Marchoux, Simond, and Salimbieri to Rio de Janeiro to study yellow fever between 1903 and 1905.

[41]Institutes of tropical medicine were founded in London (1899), Liverpool (1899), Hamburg (1900), and Brussels (1906).

[42]For a description of the Pasteur Institutes established abroad by the French colonial government, see Pasteur Vallery-Radot, "Les Instituts Pasteur d'outre mer," *La Presse Médicale* 21 (Mars 1939), 410–413; P.-Noel Bernard (*ed.*), *Les Instituts Pasteur d'Indochine (Centenaire de Louis Pasteur, 1822–1895*) (Saigon: Imprimerie Nouvelle Albert Portail, 1922); A. Calmette, "Pasteur et les Instituts Pasteur," *Revue d'Hygiene* 45 (1923), 385; and A. Calmette, "Les missions scientifiques de l'Institut Pasteur et l'expression coloniale de la France," *Revue scientifique* 89 (1912), 129.

[43]See the next chapter for an account of this episode in the history of the Oswaldo Cruz Institute.

[44]See Chapter 6 for a further discussion of the role of foreign experts in science in developing countries.

[45]The imitation of the German university, and its adaptation and transformation in the political and educational context of America in the 1870s and 1880s is very well described in Laurence R. Veysey, *The Emergence of the American University* (Chicago: University of Chicago Press, Phoenix Books, 1970), pp.121–179.

[46]See, for instance, Marcel Roche's account of the development of science in Venezuela, where he makes a similar point. Marcel Roche, "Social Aspects of Science in a Developing Country," *Impact of Science on Society* 16 (1966), 51–60.

[47]George Basalla, "The Spread of Western Science," *Science* 156 (5 May 1967), 613–617.

[48]See Chapter 7 for further reference to the Butantã Institute.

5

From Serum Therapy to Research: Science and Politics in Brazil, 1903-1908

In 1902 the Paulista statesman Francisco Rodrigues Alves was elected President of the Republic of Brazil. Rodrigues Alves was a well-known politician with several years of experience in state and national office behind him. Especially important for the future of science were his two years as governor of the state of São Paulo between 1900 and 1902, where one of his tasks had been to remodel the state's public health services in order to serve the needs of the rapidly expanding economy and population.[1]

Of the public health problems in the state capital of São Paulo itself, yellow fever was not the most pressing. The high altitude of the city prevented the *Aedes aegypti* mosquito from breeding and therefore from infecting the city population. Yellow fever was, however, epidemic in the port of Santos, and in the small towns serving the coffee plantations, such as Sorocaba and Ribeirão Prêto. With coffee at the heart of the state economy, the question of yellow fever figured prominently, therefore, in the activities of the state's department of public health. In 1900, only a few months before the Reed Commission in Havana announced the discovery of the transmission of yellow fever by the mosquito the São Paulo physician Luís Pereira Barreto remembered receiving a note from Governor Alves commenting on a series of discussions on the causes of yellow fever that had been held at the São Paulo Medical Society.[2] When the director of public health, Dr. Emílio Ribas, made public the announcement of the Reed Commission shortly afterwards, it was Governor Alves who authorized Barreto, Ribas, and Dr. Adolfo Lutz, director of the state Bacteriological Institute, to carry out a series of inoculation experiments on human beings in order to confirm the Reed Commission's findings. Meanwhile, prophylactic measures against yellow fever epidemics based on the elimination of the *Aedes* mosquito were put into effect in the state by early 1903. The inoculation experiments, which were delayed for a variety of reasons until 1903, climaxed a decade of work in the hygiene sciences that had placed São Paulo at the forefront of the public health movement in Brazil.

Rodrigues Alves' contacts with the public health movement in São Paulo in the period immediately preceding his assumption of the presidency of Brazil were reflected in his campaign speeches. The connection between São Paulo and Rio de Janeiro in this respect is an interesting example of the role that a state government can play in stimulating new programs of administration in a federal capital. Alves campaigned for the presidency on the theme of renovation—the need to renovate Brazilian culture, Brazilian social life, and the Brazilian economy in order to make Brazil truly a part of the civilized world. Here we hear echoes of the ideas expressed in Barreto's letter to Ribas in 1899 quoted earlier—the desire to bring Brazil into the modern world and to bridge the gap between the country's potentiality for greatness and the sad reality of its diseased and backward condition.[3] By renovation Alves included the sanitation of the port of Guanabara and the capital of Rio de Janeiro. At the time of his campaign, in fact, many foreign ships bypassed Rio rather than risk infection by entering the port. The need to take action against disease was pressing not only in order to improve Brazil's public image abroad, but also in order to prevent a reduction in much-needed immigration into the country. Alves' intention to develop a more effective federal policy of public health meant that for almost the first time in Brazil's history the quality of health in the national capital was to become one test of the nation's claim to civilization.

Cruz and the Yellow Fever Campaign

With the decision to give priority to sanitation, Alves began the search for a new director of the federal department of public health.[4] Traditionally the position had been held by a medical doctor, and the department itself had been small. According to the incumbent of the post in 1902, Dr. Nuno de Andrade, the staff consisted of himself, five aides, one sanitary inspector, five servants and a coachman. His requests to the Minister of the Interior in 1902 for an expansion in staff and money in order to begin to put into effect some of the new techniques against yellow fever had been met with indifference.[5] The small size of the federal department of public health in 1902 was in part due to the relatively low priority assigned to public health at the time and the disillusionment that had followed previous attempts to deal with yellow fever. It was also the result of the structure of the state and federal public health administration. On the principle of states' rights, each state bore responsibility for the sanitation of its own territory. The federal authorities bore responsibility for the defense of the Union against invasion of disease from the outside. In effect the federal authorities directed their activities toward the imposition of quarantines in the major ports.

In the capital city of Rio de Janeiro, federal action was restricted to the federal district, while the responsibility for the sanitation of the city fell to the Rio municipality. Federal officials could only intervene in the city's business if requested to do so by the municipal authorities. The resulting inefficiency of the dual sanitary administrative structures was great.[6] Moreover, the sanitation problems of the city of Rio had already outdistanced the financial ability of the municipality to deal effectively with them. When bubonic plague had threatened to invade the city in 1899, for instance, the city's plans to fund the small laboratory at Manguinhos for the production of anti-plague vaccines and serums were disrupted by the outbreak of another disease requiring the financial resources of the municipality. As a result the laboratory was handed over to the federal authorities. A population which was already over six hundred thousand was crowded into often unsanitary houses, many situated on unpaved streets with open sewers. Not only were yellow fever, the plague, and smallpox great killers, but measles, scarlet fever, influenza, and the less dramatic but ever present intestinal diseases and tuberculosis took their annual toll of human life. A logical solution to the problem of public health in the capital lay in the centralization of public health administration under the federal government, a process of centralization that had already occurred in a large number of European countries. It was also a way, of course, of making Brazilians in all states subsidize the sanitation and beautification of the national capital.

President Rodrigues Alves sought for the post of director of public health a person with political ability, a knowledge of the hygiene sciences, and considerable administrative skill. Resistance to new sanitary legislation was a possibility, but compliance could not depend upon the use of martial law, a technique used by the American military authorities in Havana during their yellow fever campaign. The position called for a man capable of seizing the opportunity to draw up an entirely new program with new administrative mechanisms for registering, identifying, and controlling disease, and for applying new medical technology effectively.

According to Sales Guerra, Cruz' biographer, the post was first offered to Sales Guerra himself.[7] Sales Guerra refused on the grounds that the position required a skilled scientific technician.[8] Instead he recommended Oswaldo Cruz, whose career he had followed with interest, whose work in bacteriology he had encouraged, and with whom he had discussed the newest findings of the Reed Commission. The government reacted with surprise to the proposal. The Minister of Justice and the Interior, J. J. Seabra, asked "Oswaldo Cruz? Who is he?" Cruz was sent to meet the President and his key Ministers with his proposals for the position. To Alves, Cruz sketched out an ambitious plan of action against three of the major epidemic diseases of Rio which were capable of immediate control—yellow fever, smallpox, and the plague. Yellow fever was to be eliminated by the systematic extermination of *Aedes aegypti* mosquitoes and mosquito larvae, by destruction of their breeding places, and through the identification and strict isolation of yellow fever patients. Smallpox was to be dealt with by compulsory vaccination of the Rio population. Plague was to be controlled by the destruction of the grey rat in the city and the use of vaccines and serums. As Cruz envisaged it, the program would be large in scale, focused in its goals, expensive, but therefore effective.

President Rodrigues Alves and his Minister, J. J. Seabra, were impressed by Cruz as a man of education and science, whose commitment to the importance of sanitation must have been both refreshing and convincing. Alves' familiarity with the work of the Americans in Havana no doubt contributed to his acceptance of Cruz' proposal. In Alves, in fact, Cruz found a politician in national office who was virtually unique in his grasp of the significance of the modern sanitation sciences, and in his ability to give the strong executive lead a huge program in public health required. The backing of Alves and his Ministers during the public health campaign was a crucial element in the campaign's eventual success.[9]

To Oswaldo Cruz, therefore, Alves entrusted the task of bringing to an end the epidemics of yellow fever and smallpox that had done so much to give Brazil its reputation as one of the least healthy spots in the tropics. The announcement of Cruz' appointment as director of the federal Department of Public Health in March, 1903, displeased many, for Cruz at the time was young, virtually unknown, and almost without administrative experience. Many doubted that someone whose entire professional career had until then been confined to the narrow arena of the scientific laboratory would possess the necessary political and administrative skills to handle the program that Alves, under Cruz' advice, now presented to Congress in the form of a new sanitary bill for the Union.[10] Cruz' detractors forgot, in the uproar that ensued as debate opened on the controversial bill, that many of the most successful public health administrators of the period were not politicians but scientists drawn from the same background as Cruz, and with the same laboratory specialization.

As the debate got underway, Cruz took command of the department and sent a delegate to Havana to observe the work of the American authorities in Cuba. In April, the new Serviço de Prophylaxia da Febre Amarella (Yellow Fever Prophylaxis Service) was organized out of funds authorized in December of the previous year. The Service was composed of a technical director, five medical inspectors, teams of horses to carry material for destroying mosquito larvae, carpenters to construct isolation units both within homes and in the Isolation Hospital of Rio de Janeiro, and thirty-six guards to enforce compliance with the new regulations. The specific task of the Yellow Fever Prophylaxis Service was to divide the entire city into sanitary districts, to undertake sanitary policing of suspected zones of infection, to destroy foci of mosquitoes, and to identify, register, and isolate yellow fever patients. Already, by April 20, 1903, the first rigorous isolation of yellow fever patients had begun. Bulletins were prepared by the department explaining the work of the Service and the significance of the Finlay doctrine. Some of these were sent to the major newspapers so that the political elite and the reading public could be informed of the new measures.[11]

In Congress, discussion of the sanitary bill was long and often acrimonious.[12] In addition to the creation of a large-scale and costly yellow fever service based on the still somewhat controversial findings of the Reed Commission, the bill proposed to institute compulsory vaccination, to unify the federal and municipal health services for the city of Rio de Janeiro, and to create a uniform sanitary code for the entire Union.[13] Many congressmen interpreted the latter proposal as an infringement of the deeply cherished right of every state to determine its own health policy, and a dangerous step toward centralization. The creation of the yellow fever service was opposed because of its reliance on what seemed to many a mere "opinion" or unproved hypothesis concerning the mosquito and yellow fever. Still others objected to the cost of the bill (five and a half million milreis), others to Cruz' youth and inexperience.[14] Despite these criticisms, however, the bill was finally passed in a truncated form in December of 1903, the matter of compulsory vaccination having been set aside for separate consideration. Funds were formally released to pay the large numbers of technicians and sanitary inspectors recruited to the department of public health. The new regulations were put into effect, and the brigades of "mosquito killers" became a familiar sight in the city. A diagnostic laboratory was organized to carry out identifications of diseases as they were reported by the inspectors, and isolation wards were installed in hospitals.

The "yellow fever campaign," as the sanitary campaign is often called, forms one of the most fascinating chapters in the history of the Old Republic in Brazil, as well as a most interesting episode in the history of public health. The influence of the campaign on Brazil's economy, on immigration into the country, and on Brazilians' conception of themselves and their future, has yet to be estimated. The story of the campaign still awaits a major study, and it is not the purpose of this book to provide one.[15] Nonetheless, its impact on the development of science in Brazil was so important that a brief account of the campaign and of Cruz' role in it is essential.

The campaign took place in an atmosphere of hostility, publicity, and, at times, revolt. Many factors contributed to making the campaign a major political issue and a target of anger. The Positivist Church in Rio was opposed to compulsory vaccination on the grounds that it limited freedom of choice. The military positivists were especially opposed on these grounds, and for complex political reasons were determined to make trouble for the government. Commercial interests disliked the disruption to business caused by the cleaning and widening of the streets that was part of the program, and by the attention the campaign called to the unsanitary condition of the city. The public health campaign became, in short, an excuse by many different groups opposing President Rodrigues Alves to discredit the government.

Cruz himself was made the butt of newspaper jokes, which published caricatures of him as a monster ruthlessly imposing the cruel techniques of science upon a cowering population.[16] His name became a household word. The more conservative members of the medical profession used the popular press to express their dissent from the Finlay doctrine and to charge Cruz with delinquency for not continuing to use the time-honored practice of fumigation and disinfection in the treatment of yellow fever cases. Cruz' predecessor in the directorship of public health, Dr. Nuno de Andrade, went so far as to forget his advocacy of the Finlay doctrine in 1902 and to publish a series of articles in the prestigious Rio newspaper, the *Jornal do Commercio*, in the summer of 1903, in which he argued that until the pathogenic agent of yellow fever itself was isolated, the Finlay theory was not a proven fact.[17]

The public resisted the entry of the sanitary inspectors because they were uncertain of the purpose of many of the sanitary measures, because they were alienated from the government, and because they were fearful of what would happen to them. Indeed, special tribunals were established to force compliance with the new sanitary regulations, and the sanitary campaign was pursued relentlessly. Resistance to compulsory vaccination was so great, however, that the part of the legislative bill dealing with it was separated from the main bill in 1903, in order not to jeopardize the passage of the bill bringing about the unification of the sanitary services and the construction of a uniform sanitary code. Compulsory vaccination finally became law in October of 1904.[18] By this time opposition among the military positivists was so intense that the Military Academy spearheaded a military revolt which for a short while threatened to bring about the collapse of Alves' administration. Cruz offered his resignation, which was refused by Alves, and the opposition was finally crushed and order restored. Nonetheless, the law failed to be put into effect and a large part of the population in Rio de Janeiro remained unvaccinated. In 1908, while Cruz was still director of public health under a new government, the city paid the consequences when it experienced one of the worst outbreaks of smallpox in its history. Over nine thousand died from smallpox alone that year.[19]

This was President Rodrigues Alves' only major defeat in the area of sanitation, however. By the terms of the bill of 1903, Cruz had been given just under three years to eliminate yellow fever in the capital. If, by the end of 1906, when Rodrigues Alves' presidential term expired, yellow fever was still epidemic in the city, the new Congress would be given the authority to revoke the sanitary law. In 1906, however, Cruz was able to report to the incoming President Afonso Pena that "yellow fever no longer exists in epidemic form in Rio."[20] In 1908, a total of four deaths from yellow fever were reported in a population that by now exceeded eight hundred thousand. In 1909 there were none.[21] The mortality coefficient of the city, which had been 25.77 per thousand when Cruz took office in 1903, had fallen to 19.97 in 1907, owing to the combined work of Alves, in undertaking a general sanitation and reconstruction of the streets and the port, and Cruz, in attacking yellow fever and the plague.

The chief significance of the campaign was the elimination of a major health hazard in the capital. The importance of the campaign for the history of science was to raise Cruz from a position of relative obscurity to one of national prominence. For almost the first time in Brazil's history, a professional scientist was widely known to the Rio public. The subject of science had been given a wide hearing in the Rio newspapers, and in the short space of three years more people had been exposed to the ideas and techniques of scientific medicine than in all of Brazil's previous history. By the end of 1906, the campaign was obviously a success, and, internationally, confidence in Brazil was on the rise. A cartoon of Cruz in 1908, the year before he resigned his post as director of public health, showed him now with a hero's wreath, and bore the caption: "This is the way the country receives its sons that honor and love the nation: it crowns and blesses them."[22]

The Congressional Debate on Science

Meanwhile, what had been the effect of the public health campaign on the status of the Serum Therapy Institute that Cruz had continued to direct during the campaign? The fact was that the years of the sanitation campaign were years of growth for the Serum Therapy Institute. Instead of a staff of three or four students, the institute already possessed a small group of first-rate medical doctors and was attracting new scientists and students. The ramshackle buildings of 1902 and 1903 were in the process of being replaced by modern laboratories. In retrospect, the key to this transformation lay in the public health campaign. Yet when Cruz came to public office, this transformation had not formed a part of the new federal policy toward science.

When Cruz submitted his proposals for a sanitary program to President Rodrigues Alves in May of 1903, he also proposed expanding the Serum Therapy Institute, which was technically an adjunct of the federal public health department, into a center "for the study of infectious and tropical diseases along the lines of the Pasteur Institute of Paris."[23] The new institute, he argued, should be charged with the teaching of bacteriology, the production of vaccines and serums, and original research. It was in order to achieve these ends that Alves included in the new sanitary legislation of 1903 a plan for the reorganization of Manguinhos, through its separation from the public health department, and its formation into a single, independent institution of science under the direct authority of the Minister of the Interior and Justice. This step would give the Institute equal status with the Medical Schools of Rio de Janeiro and Bahia. Cruz also requested a tripling of the Institute's annual budget, as well as the establishment of a permanent fund to endow research, the fund to be made up of government bonds and therefore not subject to annual congressional accounting.[24] In this way it was hoped to protect research funding from year-by-year changes. Cruz' request to make Manguinhos a teaching institution, however, was not included in the legislation sent to Congress in June for debate.

The congressional debate on the proposal for the Serum Therapy Institute was inextricably bound up with the debate on the sanitation bill, and was hostile. Few of the members of Congress were willing to alter the character of what seemed to many a practical laboratory already serving the immediate needs of the public, in order to make it into a center of science at great cost, merely, as one Deputy hinted, to satisfy the youthful ambitions of its director. The congressional committee established to analyze the matter of the institute was divided in its opinion. One member of the committee argued that Cruz meant to profit financially by holding two jobs, one as director of public health and the other as director of the proposed institute. Cruz' plan to separate the institute from the department of public health was viewed as a cynical plan to evade public scrutiny. Mr. Brício Filho commented that it would leave Cruz in the "enviable" situation of having "many privileges," a life tenure without "superiors" to supervise him, in contrast to other federal employees.[25] Another termed Cruz' suggestions a "grave error, an offense in the face of the economic situation," "an unique exception and a hateful one, an attack on the constitution, a dreadful and dangerous precedent."[26] These remarks were greeted with applause and cries of "Hear! Hear!" It was clear that few of the legislators shared Cruz' belief that public health could not improve without a major change in the institutional organization of the biomedical sciences.

The discussion continued acrimonously over the weeks. Cruz, however, had his defenders. Deputy Mello Mattos, who had placed the bill on the floor for Cruz, pointed out in reply to the bill's critics in October that Cruz was relatively well-off and had no need to gain financially by holding two posts. He characterized Cruz as a dedicated scientist, motivated by a genuine desire to aid the country, as a man who had chosen to leave behind the chance of a scientific career in Europe for an uncertain career in Brazil, and a man whose scientific credentials were such that any public health program placed in his hands was guaranteed of success. He also explained the reasons why Cruz asked for financial and administrative independence for Manguinhos, summing up succinctly for his listeners some of the problems associated with science in a developing country:

> It is indispensable that in a purely scientific establishment, the scientific work should be, whenever possible, always the same, free of the encumbrances and changes that normally affect dependent institutions during changes of administration; for this reason the Serum Therapy Institute cannot and should not be subordinated to the department of public health, whose scientific orientation can change as often as does the director. . . .In the administration of the institute, the institute would still remain a dependency of the Minister of the Interior, who is placed higher in the administrative structure than the director [of Public Health], and who would have the authority to dismiss the director of the institute. It is not true, therefore, to say that the director of the institute would have a lifelong post and be outside the control of the government.[27]

As for the desire for financial independence expressed in the request for a permanent fund of government bonds, a scientific institute, Mello Mattos argued, could not be at the mercy of the caprices and hardships of the government of the day. It had to possess a secure budget reserved for whatever needs the institute had. However, since this matter aroused particular opposition, Mattos said he was willing to drop the question of financial autonomy via the government patrimony if the rest of the bill could be accepted. As a result of this move, in fact, amendments were added to the proposal removing both the administrative and financial independence requested by Cruz. His proposal that the director of Manguinhos be authorized to contract foreign scientists to work at the Institute was also eliminated.[28]

There were others who also showed some understanding of the possible importance of the institute envisaged by Cruz for Brazil. One deputy remarked that because Brazilians were on the whole ignorant of medical science, they assailed scientific proof [of the transmission of yellow fever by the *A. aegypti* mosquito], and ideas based on careful observation, with an "insolent audacity," trying to refute scientific discoveries with "ingeniously simple" arguments.[29] José Bonifácio, a distinguished member of the Commission on Public Health and Education in Congress, rose to speak on October 27th, 1903, after discussion had continued over a period of several weeks. "Rio de Janeiro," he remarked, "is the capital of the Republic, the seat of the federal government, and the great center where all the concerns of the states and the products of their riches converge:"

> As long as we do nothing to look after our health conditions, undertake sanitary programs, prevent the invasion of contagious disease . . . we have to acknowledge the sad fact that abroad, in international congresses, our beautiful city, so full of natural beauty, is pointed out as one of the most insalubrious of places, where regularly appear diseases that could be prevented, infections that could be suppressed, and ills that could be eliminated but are not eliminated.[30]

Applause followed his speech. Through support of José Bonifácio and others, and the realization of the urgency of the sanitary situation, the sanitary bill was finally passed in December and the money authorized for the yellow fever service.

Cruz' plan to create an independent center of experimental medicine that would carry out research in the microbiological field and produce vaccines and serums did not fare so well. On the final vote, almost all of his proposed changes for Manguinhos were vetoed. Neither administrative nor financial autonomy were granted.[31] The small budget was, however, doubled, from 60,000 milreis to 120,000 milreis (approximately 30,000 dollars, at 0.25 dollars a milreis) and a small additional sum granted for the improvement of the buildings. An advantage was also gained in that the Serum Therapy Institute, along with a newly established diagnostic laboratory in the public health department, could be considered the agencies by which some of the tasks specified in the new sanitation bill would be carried out, namely the study of

> the nature, etiology, treatment and prophylaxis of transmissible diseases that appear or develop in any part of the Republic, where resources and services organized for research of a technical or scientific kind do not exist, and are deemed necessary, or where they do exist but the government deems it convenient to undertake such studies.[32]

The congressional rebuff to Cruz' plans for Manguinhos perhaps demonstrated the legislator's shortsightedness, but was not surprising given Cruz' lack of political reputation in 1903. The decision did not deter Cruz. As the newly appointed director of public health, Cruz was in charge of a large, well-funded program, and the Serum Therapy Institute was incorporated fully into the program as time passed. The technicians appointed as sanitary inspectors were directed to Manguinhos for training and to help in the preparation of serums and vaccines. Medical students from the Rio Medical School sought employment in the sanitation program, and they too found their way to Manguinhos to prepare their medical theses.[33] Word got back to the city that at Manguinhos "a new kind of science was being practiced," and that "it was not therefore very surprising that new people, the curious and the interested, were attracted there. A real pilgrimage of students and doctors came to see the Institute and to Cruz, all eager to begin this work," in the words of one of those who made the pilgrimage and stayed on as a staff member.[34] As the human resources grew, so did the material resources. Microscopes, glassware, and experimental animals began to be supplied in increasing quantities. Books and journals were bought for a new library collection, and the work of the students published bearing the imprint of Manguinhos.[35]

These changes at Manguinhos were not kept a secret. In his official report to the Minister of the Interior for 1905, Cruz commented that the institute, created for the extremely limited end of preparing serums and vaccines against bubonic plague,

> was rapidly expanded in its functions until it has become an institution analagous to the type found in all civilized countries . . . . that is, it hopes to be an institute in which is studied from all points of view infectious diseases, principally those that most afflict our own country.

He pointed out in the same report that while the facilities were inferior to those of similar institutions in Europe, the same was not true of its work. Its anti-plague serum had already been tested by the Pasteur Institute in Paris and found to be as good as, if not in some ways better than, those produced elsewhere. He mentioned the preparation of new serums at the institute, such as anti-diphtheria serum, which were not yet ready for public sale, and listed some of the studies being carried out by the students and staff, and the medical theses prepared at Manguinhos and already presented to the Rio Medical School. He brought to the attention of the authorities the fact that the buildings at Manguinhos had originated as a "little country cottage of one story high, built to house an engineer in charge of an incinerator works," and added:

> It is unbelievable that, in this minute house, which is no doubt unmatched in its poverty even by the most useless of provincial laboratories, we are trying to do the same work as that being carried out in the large and comfortable European or American institutions. The real wastage and sacrifice this effort entails, and the danger, can only be appreciated by those who know how an institute such as ours works.[36]

In order to remedy this situation before the Alves administration came to an end in 1906, Cruz commenced a project to construct new laboratory facilities in 1905. A Portuguese architect, Luis de Morães, was contracted sometime in 1905 to plan the facilities; a rough sketch by Cruz, now in the *Oswaldo Cruz Files*, and bearing a date of 1905, provided the outline on which the architect worked.[37] Marble was imported from Portugal to provide the intricate tiling that gives the distinct character to the Oswaldo Cruz Institute as it exists today. Materials for construction were unloaded at the small dock at Manguinhos and hauled up the hill by horses. Work began at the site in early 1906, and prompted a congressional inquiry, but while Congress delayed authorization Cruz continued the construction, giving physical reality to the "Pasteur Institute" he had in effect begun to organize at Manguinhos.[38] Eventually, the finished buildings housed some of the most efficient medical laboratories in South America. Architecturally somewhat bizarre yet imposing, the buildings stood on the hill overlooking the wide plain separating Manguinhos and the city, and overlooking the bay of Guanabara. In its patterned tiling and Moorish cupolas, the institute bore a strong resemblance to the Meteorological Observatory of Montsouris in Paris, and Dr. Olympio da Fonseca, a former staff member of the Institute, suggests Cruz knew the Observatory from his stay in Paris, since the great authority on the bacteriology of air, water, and earth, Antonin Pierre Miquel, worked there.[39] Cruz knew the work of Miquel well, even before Cruz' studies in Paris, and cited his works in his medical thesis a number of times.[40]

The Founding of the Oswaldo Cruz Institute

In 1906, Rodrigues Alves' term as president ended and his successor, Afonso Pena, prepared to take office. The yellow fever campaign had proven itself a success and plague had been virtually eradicated. Cruz was a national hero, personifying to many the triumph of truth over ignorance, of science over superstition. At the height, therefore, of his political prestige, and because no doubt he wished to consolidate the position of the Serum Therapy Institute before a change in government, Cruz once more presented a bill to Congress, through Deputy Mello Mattos. This bill, virtually identical to the bill of 1903, asked for the legal creation of a "Pasteur" Institute in Brazil.

Now, after the success of the sanitation campaign, Congress found Cruz hard to refuse. After debate in the house and senate, the institute was renamed the Institute of Experimental Pathology of Manguinhos and its budget tripled. A further sum of 600,000 milreis was voted to complete construction of the new laboratory facilities. The institute was severed from the department of public health, and organized as an independent institution of science under the Ministry of Justice and the Interior. The director was to be elected by a special, technical commission and appointed by decree. Financial independence was established through a permanent fund of government bonds. The institute was authorized not only to prepare vaccines and serums, but to study infectious and parasitic diseases, to create a veterinary school and to organize scientific commissions of inquiry. A scientific publication, the *Memorias do Instituto Oswaldo Cruz*, was authorized. The institute was given the right to offer contracts to scientists to work on its staff, and the staff itself was formally divided into six departments. Eight staff members, excluding Cruz as director, were appointed to permanent positions by Cruz, chosen from among the technicians and doctors already working at the institute. The passage of the bill in 1907 represented a personal victory for Cruz, and a major step in the advancement of science in Brazil.[41]

A final stage in securing the public position of the institute occurred in 1907, when Brazil was invited to participate in the XII International Conference of Hygiene being held in Berlin. Brazil was the only country in Latin America to receive an invitation, and the invitation was the result in part of the presence of Dr. Henrique da Rocha Lima, a member of the institute, in Berlin at the time. Cruz determined to use the opportunity in Berlin to demonstrate the quality of Brazilian science, and while Congress delayed in voting funds to send the scientists to the conference, he and his staff went to work to prepare exhibits describing the functions and activities of the institute.[42] Models were erected to demonstrate the future physical facilities of the laboratories, and Cruz drew up a brief synopsis of the history of the yellow fever campaign in Rio de Janeiro. The techniques used to isolate yellow fever patients in their homes and in the Isolation Hospital were illustrated. Dr. Artur Neiva described the studies being made of new species of mosquitoes in Brazil, and Dr. Carlos Chagas evaluated the recent anti-malaria campaigns organized by the staff members in the state of Rio de Janeiro. What most excited the attention of the other scientists attending the conference was the announcement by Dr. Henrique de Beaurepaire Aragão that he had unraveled the life-cycle of the plasmodium in the tissues of the common pigeon. This discovery was extremely important when the life-cycle of the malaria plasmodium in humans was studied later.[43]

For its contributions to the advance of the hygiene sciences, the judges at Berlin awarded the Institute of Experimental Pathology at Manguinhos its highest prize, the gold medal. Brazilians in the capital responded to this news with euphoria and Cruz was practically mobbed in the port of Guanabara when he arrived back in Rio. The following spring, a presidential executive order renamed the institute the Oswaldo Cruz Institute, in honor of Cruz' services to Brazil.[44] Success in Berlin also brought to a conclusion Cruz' campaign to secure research funds by allowing the institute to sell the vaccines and serums it produced for profit.[45] In 1906 this had been one of the few requests denied to Cruz. Following the conference at Berlin, permission was granted to market serums and vaccines, and in later years this resulted in an increase in the institute's income of as much as one-third.[46]

Several things stand out in the development of the Serum Therapy Institute into the Oswaldo Cruz Institute between 1903 and 1908. The initial "crisis" period from 1899 until 1902 had been followed by an explosive period of growth that was closely tied to, if not almost entirely dependent upon, political events. The highly public campaign against disease gave Cruz a visibility and legitimacy within the political structure that was almost unprecedented for a scientist, and suggests the sensitivity of scientific change to politics in countries where science is not routinely funded and supported. That Cruz used his political power to establish a research institute was an unusual departure from tradition in Brazil, and was the result of his professional training in the microbiological sciences, which led him to attempt to secure the conditions for a professional career within his own country. To carry out this task, however, meant a temporary abandonment of the role of laboratory scientist for that of scientific entrepreneur and administrator. Although Cruz was already a member of the scientific and political elite in Brazil, his career as a scientific entrepreneur represented a distinct innovation. The yellow fever campaign provided the right political opportunities for the right professional scientist to establish new conditions for science in Brazil. The fact that the microbiological sciences had an immediate practical effect aided the rapid institutionalization of these sciences.

It was not surprising that the explosive institutional growth occurred independently of the existing scientific and medical institutions. Ben-David has described how the rise of semi-autonomous research institutes as a major locus of scientific activity allowed support to be given to such institutions regardless of the degree of support for science among the public at large.[47] The Oswaldo Cruz Institute clearly exemplified this pattern of growth. The defeat in

1903 of Cruz' first proposal to make the institute a center of experimental medicine had had unexpected benefits. Instead of new and restrictive legislation imposed by non-scientists, the institute, as an adjunct of the sanitation program, was given freedom to expand its size and range of activities. Cruz also profited by the political events of 1903–1906 and increased the institute's budget and facilities. While the time was not ripe for a wholesale re-evaluation of the role of science in the Brazilian economy, the independent evolution of a specific institution that had demonstrated its ability to be successful offered a rewarding route to scientific development. This, at any rate, appears to be one of the lessons to be drawn from the Oswaldo Cruz Institute's history between 1903 and 1908.

In 1909, when Cruz resigned from his post as director of public health to devote his energies entirely to the development of the institute, several problems awaited solution despite the high esteem Cruz enjoyed. There was a danger that power within the institute would be concentrated too exclusively in Cruz' hands, and some administrative mechanism had to evolve to ensure the institute's survival of Cruz' eventual resignation or death. Expansion of a research and applied science program depended on the continuous recruitment of trained scientists if the institute were to live up to its designation as a center of experimental medicine. The shortage of trained scientists in Brazil raised afresh the possible role of foreign microbiologists at the institute. The isolation of the institute from the other medical and scientific institutions in Brazil also raised the question of whether it would remain a unique example of active science, or would influence the evolution of other institutions. The continued support of successive governments, once Cruz was out of national office and the excitement generated by the yellow fever campaign died down, depended upon consolidating a working relationship with the government. The move towards research and away from applied science would test this relationship. In short, as Claire Nader points out in a perceptive analysis of science in developing countries today, what was called for was "continuity in scientific enterprises from policy to research to application" requiring "long-term commitment" and "protection from political instability."[48] The methods by which some of these fundamental tasks of scientific institution-building were tackled by the Oswaldo Cruz Institute are the subject of the next chapter.

References

[1]President Rodrigues Alves lacks a full-scale biography. Details on his political career and his government have been taken from the following sources: Francisco de Paula Rodrigues Alves Filho e Oscar Rodrigues Alves (*eds.*), *Centenário do Conselheiro Rodrigues Alves*, 2 vols. (São Paulo: Empresa Gráfica da "Revista dos Tribunais" Ltda., 1951); José Maria Bello, *A History of Modern Brazil, 1889-1964* (Stanford, California: Stanford University Press, 1966), Chapter 14; and João Pandia Calógeras, *A History of Brazil* (New York: Russell and Russell, Inc, 1963), Chapter 14.

[2]According to Dr. Pereira Barreto, Alves sent a letter to Barreto at the São Paulo Medical Society congratulating Barreto for his statement on the mosquito theory of the transmission of yellow fever. See Francisco de Paula Rodrigues Alves and Oscar Rodrigues Alves, *op. cit.*, Vol. 1, pp. 78-80.

[3]See Chapter 3, reference 34.

[4]Technically speaking, the position of director of public health had become vacant owing to the Alves government's rigid application of the law against the accumulation of jobs, which prohibited anyone from holding more than one public office.

[5]Brazil, Directoria Geral de Saúde Pública, Placido Barboso e Cassio Barboso de Rezende, *Os serviços de saúde pública no Brasil, especialmente na cidade do Rio de Janeiro de 1808 à 1907* (*Esboço histórico e legislação*), 2 vols. (Rio de Janeiro: Imprensa Nacional, 1909), Vol. 1, pp. 107-108.

[6]*Ibid.*, Chapter 4, gives a good account of the federal and municipal health services in the period before 1903. Inefficiency was also related to *empreguismo*, the use of patronage in awarding jobs. The prefect of the federal district was appointed by the President of the republic.

[7]It is not clear exactly why Sales Guerra was offered the position. It appears he was a friend of the new Minister of Justice and the Interior, J. J. Seabra, and had discussed with him the difficult task of finding a replacement for the outgoing director. See E. Sales Guerra, *Osvaldo Cruz* (Rio de Janeiro: Casa Editôra Vecchi Limitada, 1940), pp. 58-59.

[8]*Ibid.*, p. 59. His refusal may also have been motivated by the fact that he did not wish to offend the outgoing director, Dr. Nuno de Andrade.

[9]The Cabinet included J. J. Seabra as Minister of Justice and the Interior, Lauro Müller, an Army major with an exclusively non-military career, as Minister of Transport and Public Works, Leopoldo de Bulhões as Minister of Finance, and Baron Rio Branco as Minister of Foreign Affairs. José Maria Bello comments that Alves was extremely astute in selecting his aides; Alves reportedly made the comment, "My ministers do anything they wish, except what I don't want them to." See José Maria Bello, *op. cit.*, p. 175.

[10]This point was made by Dr. Ezequiel Caetano Dias, a staff member of the Oswaldo Cruz Institute, in his *Traços biográficos de Oswaldo Cruz* (Rio de Janeiro: Imprensa Nacional, 1945), p. 17.

[11]For the yellow fever campaign, I have used, in addition to E. Sales Guerra's biography

of Cruz, the two-volume account of the history of the public health services in Brazil and the yellow fever campaign put together by Placido Barbosa and Cassio Barbosa de Rezende, under the auspices of Oswaldo Cruz in 1909. These massive volumes reprint in full much of the major legislation for public health from 1808 to 1909. See Brazil, Directoria Geral de Saúde Pública, *Os serviços de saúde pública no Brasil*, 2 vols. Also useful are: Octavio G. de Oliveira, *Oswaldo Cruz e suas atividades na direção da saúde pública brasileira* (Rio de Janeiro: Servico Gráfico do Instituto Brasileiro de Geografia e Estatística, 1955); and Theophilo Torres, *La campagne sanitaire au Brésil* (Rio de Janeiro: Direction Générale de la Santé Publique, 1913).

[12]The Congressional debates on the sanitary program presented by President Rodrigues Alves are found in Brazil, Congresso Nacional, *Annaes da Câmara dos Deputados*, 1903, Vols. 1, 3, 5, 6, 7 and 8.

[13]The sanitary code is reprinted in its final form in Brazil, Directoria Geral de Saúde Pública, *Os serviços de saúde pública no Brasil*, Vol. 2, pp. 892–1011, as Decreto N. 1, 151 de 5 de janeiro de 1904, *Reorganização dos serviços de hygiene administrativa da União.*

[14]See the 1903 volume of the *Annaes da Câmara dos Deputados* cited in reference 12.

[15]Donald B. Cooper, in "Oswaldo Cruz and the Impact of Yellow Fever on Brazilian History," *The Bulletin of the Tulane University Medical Faculty* 26 (February 1967), 49–52, touches on some of the consequences of yellow fever and its final elimination on Brazilian history.

[16]All important politicians and administrators were so treated in the weekly papers. The Oswaldo Cruz Filho family possesses an extensive collection of articles and caricatures taken from the newspapers during the public health campaign. See the bibliography for a description of this source. I wish to thank Dr. Oswaldo Cruz for his generosity in letting me study this collection.

[17]Professor Nuno de Andrade's articles were republished in booklet form as *Febre amarella e o mosquito* (Rio de Janeiro: Jornal do Commercio, 1903). The publication was brought about by the resolve of two other prominent Rio medical professors, Dr. Rocha Faria, Cruz' former professor of hygiene in the Rio Medical School, and Dr. Benício de Abreu, professor of clinical medicine.

[18]This was Lei n. 1261 de 31 de outubro, reprinted in Brazil, Directoria Geral de Saúde Pública, *Os serviços de saúde pública no Brasil*, Vol. 1, pp. 163–164.

[19]Brazil, Directoria Geral de Saúde Pública, *Annuário de estatística demographo-sanitária*, 1908, p. 79.

[20]Brazil, Directoria Geral de Saúde Pública, *Relatório apresentado ao Exm. Sr. Dr. J. J. Seabra, Ministro da Justiça e Negocios Interiores pelo Dr. Oswaldo Gonçalves Cruz*, 1907, p. 5.

[21]Brazil, Directoria Geral de Saúde Pública, *Annuário de estatística demographo-sanitária*, 1908, p. 77 and 1909, p. 71. In 1907, according to the same source, New York, another city with a large immigrant population had a mortality coefficient of 18.5. In 1908, Rio de Janeiro's mortality coefficient rose to 32.48, owing to the outbreak of smallpox referred to earlier. Although generally speaking public health statistics from this period need be treated with caution, Oswaldo Cruz was responsible for inaugurating a new program of collecting data, and on the whole the statistics gathered during his tenure of the directorship of public health are reliable.

[22]In the Oswaldo Cruz Filho family files.

[23]This proposal was printed, along with the President's message, in Brazil, Congresso Nacional, *Annaes da Câmara dos Deputados*, 1903, Vol. 5, Sessões de 1 à 30 de setembro, p. 586.

[24]The details of the legislation covering Manguinhos appeared in the new sanitary legislation presented to Congress in July. See Brazil, Congresso Nacional, *Annaes da Câmara dos Deputados*, 1903, Vol. 3, Sessões de 1 à 3 de julho, pp. 17-18.

[25]Brazil, Congresso Nacional, *Annaes da Câmara dôs Deputados*, 1903, Vol. 5, Sessões de 1 à 30 de setembro, p. 578.

[26]*Ibid.*, p. 762.

[27]Brazil, Congresso Nacional, *Annaes da Câmara dos Deputados*, 1903, Vol. 6, Sessões de 1 à 31 de outubro, p. 166.

[28]*Ibid.*, p. 168.

[29]*Ibid.*, pp. 467–468.

[30]*Ibid.*, p. 764.

[31]The final debate and vote on the amended bill occurred in December. See Brazil, Congresso Nacional, *Annaes da Câmara dos Deputados*, 1903, Vol. 8, Sessões de 1 à 29 de dezembro, p. 93. The project became law on January 5, 1904.

[32]Article I, Section Ia, of *Regulamento dos serviços sanitários a cargo da União, à que se refere o decreto n. 5,156* (1904), in Brazil, Directoria Geral de Saúde Pública, *Os serviços de saúde pública no Brasil*, Vol. 2, p. 899.

[33]There are several accounts of the early period of the Oswaldo Cruz Institute written by members of the Institute, all of them valuable for their documentation of the effect of the public health program. Here I have drawn upon the following: Henrique de Beaurepaire Aragão, *Notícia histórica sôbre a fundação do Instituto Oswaldo Cruz (Instituto de Manguinhos)* (Rio de Janeiro: Serviço Gráfico do Instituto Brasileiro de Geografia e Estatística, 1950), and his *Oswaldo Cruz e a escola de Manguinhos.* Conferência realizada no Centro Acadêmico Oswaldo Cruz, de São Paulo em 20 de septembro de 1940. Segunda edição (Rio de Janeiro: Imprensa Nacional, 1945); Ezequiel Caetano Dias, *Traços biográficos de Oswaldo Cruz* and his *O Instituto Oswaldo Cruz: resumo histórico, 1899–1918* (Rio de Janeiro: Manguinhos, 1918).

[34]Quotation from one of the ablest young scientists at the Institute, Henrique de Beaurepaire Aragão, in his *Notícia histórica sôbre a fundação do Instituto Oswaldo Cruz*, *op. cit.*, p. 12.

[35]As early as 1901–1902 more than a dozen students were working at Manguinhos, preparing their theses for graduation from the Rio Medical School. *Ibid.*, p. 13, lists some of the subjects on which these theses were written. Aragão comments that the theses had a great success.

[36]See Brazil, Directoria Geral de Saúde Pública, *Relatório apresentado ao Exmo. Sr. Dr. J. J. Seabra, Ministro da Justiça e Negocios Interiores, pelo Dr. Oswaldo Gonçalves Cruz*, 1905, p. 9.

[37]Instituto Oswaldo Cruz, *Oswaldo Cruz Files, File No. 2*, also contains photocopies of

some of the original plans for the building made by Luis de Moraes and signed by Oswaldo Cruz on July 31, 1905.

[38]An account of the diversion of federal funds from the department of public health for this project appears in Henrique de Beaurepaire Aragão, *Oswaldo Cruz e a escola de Manguinhos, op. cit.,* pp. 18–19. Since Cruz was in fact director of public health, he possessed some latitude in allocating funds, which seems to explain why the protests against his actions were not greater.

[39]I wish to thank Professor Olympio da Fonseca Filho for pointing this similarity out to me. Professor Olympio da Fonseca recently published this observation in Edgard de Cerqueira Falcão, *Oswaldo Cruz; monumenta histórica*, 3 vols., (São Paulo: Brasiliensia Documenta 6, 1971–73), Vol. 2, "A Escola de Manguinhos," p. 11.

[40]*Ibid.*, p. 11.

[41]The proposal is given in full in Brazil, Congresso Nacional, *Annaes da Câmara dos Deputados*, 1906, Vol. 8, pp. 114–118. The bill became law in 1907.

[42]A sum of 70,000 milreis was finally voted by Congress to send Oswaldo Cruz and a team from the institute to Berlin. Brazil, Congresso Nacíonal, *Annaes da Câmara dos Deputados*, 1907, Sessões de 1 a 19 de julho de 1907, Vol. 3, pp. 108–110.

[43]E. Sales Guerra, *op. cit.*, Chapter 17, describes the exhibition in some detail. For Aragão's work, see Chapter 6 of this book, section *Research Science at the Institute.*

[44]By Decree No. 6891, 19 March 1908. By article 3 of the same bill the government also authorized teaching of students at the institute. The new law, naming the institute after Oswaldo Cruz, was announced by Presidential decree in December 1907, and went into effect in March of 1908.

[45]See Brazil, *Colecção das leis da República dos Estados Unidos do Brasil de 1908*, Vol. 1, *Segunda parte*, pp. 202–208.

[46]In 1909 and 1910, for instance an income of 181,000 milreis was gained through the sale of the Institute's products. See Instituto Oswaldo Cruz, *Museum Document. Fornecimento dos productos em 1909 e 1910.* Of this sum one third came from the sale of the vaccine against the cattle disease *mal de ano*, developed at the Institute. This disease sometimes destroyed up to fifty percent of affected Brazilian and Argentinian herds. Later, the income from the vaccine rose to one-third of the total budget of the Institute. See Belisário Penna, *Oswaldo Cruz, impressões de um discípulo* (Rio de Janeiro: Revista dos Tribunaes, 1922), p. 118.

[47]Joseph Ben-David, *The Scientist's Role in Society. A Comparative Study* (Englewood Cliffs, New Jersey: Prentice-Hall, Inc., Foundations of Modern Sociology Series, 1971), especially Chapter 7.

[48]Claire Nader, "Technical Experts in Developing Countries," in Claire Nader and A. B. Zahlan (*eds.*), *Science and Technology in Developing Countries* (Cambridge: Cambridge University Press, 1969), p. 461.

6

The Survival of Science in a Developing Country: Students, Clients and Research

Between 1908 and 1920 the history of the Oswaldo Cruz Institute was a history not merely of survival, but survival as a productive institution of basic and applied sciences. The date 1920 is somewhat arbitrary, but indicates the opening of a new phase of institutional growth under a new director, and thus the institute's weathering of the death of its founder. Since science is a continuous activity demanding long-range planning and commitments, a study of why the institute survived throws light on several of the interlocking problems facing science institutions in developing countries.

At the beginning of the period under discussion, the Oswaldo Cruz Institute was in danger of being a research institute in name only. Both within and without Brazil it was known primarily for its work with the yellow fever campaign, a campaign which had not involved the development of new scientific knowledge but which had instead only applied ideas and techniques developed abroad. The granting of a new organization and budget in 1907–1908 could not in itself guarantee a flexible relationship with the government bureacracy, the definition of an appropriate research program or the recruitment of an adequate number of trained scientists. The institute was therefore in the peculiar position of enjoying great political prestige, while yet remaining an isolated example of experimental science within the educational and scientific structure of Brazil. As a consequence, its evolution from a semi-autonomous organization of applied science into a recognized institution of basic and applied science depended on its assumption of functions that would normally have been carried out by other social, political, and educational agencies in the more scientifically advanced countries. The solving of problems in three areas of institution-building proved to be particularly important for the initiation and survival of research: (1) the recruitment and training of research scientists; (2) the creation of a client relationship with the government and other agencies that could be expected to use the scientific knowledge produced by the institute; and (3) the development of a research program that would be feasible, would meet Brazilian needs, and yet not be too closely tied to local concerns. Success in each of these areas resulted in the creation of an interlocking system, involving basic and applied science, the training and employment of scientists, and the production and use of scientific knowledge within Brazil.

Staffing the Institute

Paramount among the functions of the Oswaldo Cruz Institute was the recruiting and training of medical personnel. As shown in Chapters 2 and 3, the number of institutions of science and medicine were small by the standards of Europe and the United States. Unlike industrialized countries, Brazil possessed no national or state universities where undergraduates could receive preparation in the sciences. Nor were there many industrial incentives for research. Scientific training was limited, therefore, to the professional schools of medicine and engineering. The slow development of training programs in experimental medicine in the Medical Schools of Rio and Bahia reduced their effectiveness as suppliers of trained microbiologists. The schools did, however, offer a potential source of medical students. For various reasons, another potential source of microbiologists was unavailable to Cruz, namely, foreign scientists. The threat of yellow fever was one factor discouraging emigration of scientists from Europe to Brazil. The absence of research institutions was another. By the sanitation law of 1903, moreover, the Serum Therapy Institute, as it was then, was officially denied the authority to offer contracts to foreign scientists.

It was into this research lacuna that Cruz stepped. In a remarkably short time he built up a core of medical research personnel. A sufficiently large number of students and physicians were recruited to allow Cruz to be relatively selective in his staff and to allow him to form a critical mass of researchers, all concentrating on related fields. The fact that Cruz drew entirely upon Brazilians contributed to his success. When matched with a training program, it ensured a supply of researchers within Brazil, and for the first time created the possibility of self-sustained, indigenous growth.

Several factors came together to explain Cruz' success in this area of institution-building. First was his personal and scientific authority as a nationalist, trained at one of the great European centers, yet committed to the belief that science could contribute to Brazilian development. Second was the political authority he enjoyed as director of the department of public health, one of the most influential government agencies of the day. His ability to employ sanitary inspectors and technicians to carry out the tasks of inspecting, diagnosing, and treating cases of yellow fever, bubonic plague, and smallpox presented him with an opportunity to attract new recruits to the sanitary sciences. Third was Cruz' decision to make the Serum Therapy Institute a strategic arm of the yellow fever campaign. The very success of the campaign encouraged students to enter the sanitation sciences; at Manguinhos they enjoyed training facilities and research activities not available elsewhere. What Cruz established at Manguinhos was a system of recruitment that was linked to training, and a system of training linked to employment and research.

The various ways these interlocking aspects came together can be studied through the Administrative Records of the Oswaldo Cruz Institute, which provide in synopsis a history of the staff, their reasons for joining the institute and the terms of their employment.[1] From these records it is clear that there were at least three sources of recruits—the Rio Medical School, the public health program, and a small pool of trained physicians who had already begun establishing themselves in experimental medicine. The average recruit was the inexperienced medical student, who was drawn to Manguinhos to prepare his thesis under the direction of a European-trained specialist. Students from the Rio Medical School began to come to Manguinhos on an unofficial basis as early as 1902, before Cruz became director of either the institute or the department of public health. In 1902, his staff consisted of himself, as technical director, and three others, Ezequiel Dias, Antônio Cardoso Fontes, and Henrique de Figueiredo Vasconcellos, all students. Already, in that year the names of several other students studying or completing their medical theses under Cruz appear in the records—such as Henrique Marques de Lisbôa, Fernando Magalhães, Octávio Machado, and Maria de Toledo.[2] The following year, the pilgrimage of students to Manguinhos began, among whom was Henrique de Beaurepaire Aragão, who came to work on his thesis and stayed on to become a leading member of the staff.[3] In 1907 he published his first original research paper.[4]

Cruz' appointment as director of public health in 1903 generated the second source of recruits, by increasing his opportunities to employ technicians in the sanitation program.[5] Because the department of public health lacked adequate laboratory and diagnostic facilities, a small laboratory was organized in Rio, with a professional staff of a director and four medical assistants, and a system of training and examination for sanitary inspectors begun.[6] From among those receiving appointments in the sanitation program, a number were sent to Manguinhos to aid in diagnoses and the preparation of vaccines and serums. Two of the institute's most promising students were recruited in this way, Alcides Godoy and Artur Neiva. Godoy had been employed briefly as an auxiliary technician in 1903 when Cruz sent him to Manguinhos to study the production of vaccines against the cattle disease, *mal de ano*, which had broken out among the herds of Rio de Janeiro and in the pampas of Argentina and Uruguay. A vaccine against the disease was produced in 1906, his work representing one of the first original contributions of the Institute to practical sanitation. In 1908, when the institute was authorized to sell vaccines and serums, Godoy turned over a percentage of the sales to the institute and the money was used for research.[7] The second staff member to come from the sanitation program was Artur Neiva, who like Godoy was given a small stipend to carry out investigations at Manguinhos. Alternatively, students who had completed their medical theses at Manguinhos were sometimes appointed sanitary inspectors in the prophylactic service of the yellow fever and plague campaign. Cardoso Fontes, for example, worked as a sanitary inspector in 1904. The system of recruitment was informal yet effective, since many of the recruits were first screened for their abilities at the public health department.

The third source of microbiologists in Brazil was the group of medical scientists who, already engaged in experimental medicine, found at Manguinhos the beginnings of an institutional base for research and a chance to work cooperatively with other scientists. Their recruitment gave the institute solidity as well as contacts with scientists in Europe. Not surprisingly, two of the most influential members of the Oswaldo Cruz Institute's staff came from this group. The first was Henrique da Rocha Lima, a graduate of the Rio Medical Faculty in 1902, whose interest in experimental pathology had taken him to Germany, to the laboratory of microbiology of Martin Ficker, and to the laboratories of the Virchow Institute in Berlin. Here he studied for eighteen months.[8] Before his return he was contacted by Cruz (whom he had first met in 1900 while Rocha was a medical student) and placed in charge of the pathology work at the Institute. He immediately assumed a leading role at Manguinhos, often deputizing for Cruz during the latter's absence in Rio. With the addition of Rocha Lima, the Institute benefited from both German and French traditions of bacteriology and pathology.

The second trained medical scientist at Manguinhos was Carlos Chagas. Chagas had also graduated from the Rio Medical School, and had gone to work at the Plague Hospital of Jurajubá in the state of Rio. In 1905, he was drawn into the sanitation sciences when a dock and shipping company (Companhia Docas de Santos) asked him to survey the sanitation situation in the port of Santos and evaluate the anti-malaria program of the Company. Later Chagas took part in a campaign organized by Manguinhos against malaria in the state of Rio de Janeiro on behalf of another private business company, the Rio Light, whose extension of electric power lines had been held up owing to the high mortality from malaria among construction workers. Shortly afterwards, Chagas became an official member of the staff at Manguinhos.[9]

By the spring of 1907, when the institute was made into the Institute of Experimental Medicine, Cruz had been successful enough in his informal recruitment efforts to nominate a research staff immediately (the staff was then appointed by decree). The entire staff was selected from among the students and doctors who had been at Manguinhos in an official and semiofficial capacity since 1902. It was composed of the following: director, Oswaldo Cruz; department heads, Henrique da Rocha Lima and Henrique de Figueiredo Vasconcellos (the latter taking mainly an administrative role); and staff members, Artur Neiva, Carlos Chagas, Alcides Godoy, Antônio Cardoso Fontes, Henrique de Beaurepaire Aragão, and Ezequiel Caetano Dias.[10] Of the original eight scientists nominated, at least five—Rocha Lima, Chagas, Cardoso Fontes, Neiva, and Aragão made names for themselves in research.

It is interesting that the original staff of the institute were all Brazilians who, with the exception of Cruz and Rocha Lima, had been trained exclusively in Brazil and had not even traveled abroad. Except for Cruz, who was born in 1872, the staff were all born in 1879 or 1880, and were contemporaries to within a year or two at the Rio Medical School. The staff formed a tightly knit group and possessed a unity that, though strained at times by personal rivalries, Cruz considered the institute's greatest strength. The staff sensed that at Manguinhos they were in the vanguard of an important experiment.

Since most of the staff entered Manguinhos with relatively little background in microbiology, a training program formed a crucial part of the institute's life. According to some valuable accounts of the early history of the institute, training occurred in an informal apprenticeship system. The formal system of education Cruz had received at the Rio Medical School left him convinced that research skills required a more informal approach. Under Cruz' supervision, the most elementary skills of glassware preparation and sterilization, the techniques of serum and vaccine production, and the more advanced concepts of bacteriology and protozoology, were passed from Cruz to student to student in a collaborative and cooperative effort. During the public health campaign, Cruz used a small motor launch to travel from the city to Manguinhos at least three times a week to guide the students' activities.[11] At this stage in its development, microbiology did not depend on the use of technically sophisticated or expensive equipment, and the basic concepts could be acquired fairly rapidly by students, given good teaching. Eventually, the more routine tasks of serum and vaccine production were taken over by a technical staff, some of whom had come to the institute as janitors and were semi-illiterate. The informal creation of a separate technical staff allowed the students and doctors to carry out research. By 1909, the institute's growth made the apprenticeship system less efficient, and following the new regulations allowing the institute to commence teaching courses, apprenticeship was replaced by a formal course in microbiology, based on Emile Roux' "cours de microbie technique" at the Pasteur Institute in Paris. The course lasted eighteen months and was given each day between 12 noon and 5 P.M. It was attended by groups of twenty students and on completion of the course the students were awarded a diploma from the Rio Medical School. In effect the Oswaldo Cruz Institute, as an independent body, had expropriated this aspect of medical training in Brazil so that the supply of students to Manguinhos now became regularized. The formal course was taught by staff members, several of whom had arrived at the institute virtually ignorant of microbiology.[12] Cruz' solution to the recruitment problems of science in a developing country is a striking example of what a resourceful leader, committed to national development, can do without outside help in solving the problem of generating trained personnel in a relatively brief space of time.

The first phase of recruitment and training was followed by a second phase during which, as the international reputation of the institute grew, more students attended the courses, scientists from the institute were sent abroad to perfect their training, and, in turn, a few foreign scientists came to Brazil to study tropical diseases. Both Aragão and Neiva went abroad for further study, Aragão to Europe in 1909 and 1910, where he spent time at the Munich Zoological Institute, and Neiva to the United States and Europe in 1910 to visit museums of natural history.

Of the original staff members, there were two losses in the early years. Ezequiel Dias began to show tubercular symptoms by 1906 and was sent to Belo Horizonte in Minas Gerais, which was emerging as a center of tuberculosis treatment. There Cruz organized a small laboratory for Dias and affiliated it to the Oswaldo Cruz Institute, the first laboratory to "spin-off" from the parent organization.[13] Rocha Lima's departure in 1908 was the result of ties he had formed with German pathologists in 1902 and 1903; he left to take up, first, a position in the Institute of Pathology at the University of Jena, and later at the Institute of Tropical Medicine in Hamburg, where he later became a full professor. Rocha Lima kept in close touch with the Oswaldo Cruz Institute, and was influential in putting Brazilian scientists in touch with Germans interested in their field of work. Two German scientists, Stanislaus von Prowazek and G. Giemsa, came to the Institute in 1909 to work on problems of pathology for brief periods as a result. At a later date, Rocha Lima's former professor at Berlin, Dr. Martin Ficker, also came to Brazil to reorganize the Bacteriological Institute of São Paulo. Rocha Lima's place within the Oswaldo Cruz Institute was taken by the Brazilian pathologist Gaspar Oliveira Vianna, who had previously found a temporary base in the Mental Hospital, run by Dr. Juliano Moreira. Vianna formed an important addition to the staff, carrying out valuable studies on the pathology of yellow fever and American sleeping sickness, and on leishmaniasis, before his early death by accidental infection during an autopsy in 1914.[14] The second phase of recruitment to the institute also saw the arrival of the brilliant Brazilian protozoologist, Adolfo Lutz, who left his post as director of the Bacteriological Institute in São Paulo for a research career that was to last forty years, a career made possible in part because the Oswaldo Cruz Institute offered Lutz a supportive environment for research.[15] With Chagas, Lutz was the most important scientist at the institute.

The supportive environment was aided by two other developments. The first concerned the formation of adequate library facilities, absent in the Rio Medical School. A polyglot Dutchman, Hippólito Assueros Overmeer, was hired to supervise the acquisition and cataloging of books in the fields of medical science in which the Institute specialized. Particular attention was paid to the collection of journals where the most recent scientific work was being published. By 1911, the library contained over 10,000 volumes of books and journals, making it the largest specialized scientific collection in South America.[16] To develop a better knowledge of international science, Cruz also held weekly seminars in which staff members reviewed articles in their subspecialties.[17] A reading knowledge of German was also made a requirement for staff membership; knowledge of French was assumed. The second institutional development related to equipment and supplies such as microscopes and experimental animals.[18] Through his position as Director of Public Health, Cruz was able to oversee this aspect of institution-building. He also introduced a training program in glassblowing, in order to produce the glassware necessary for the institute's work. Eventually one of the technicians trained at Manguinhos established one of the first companies to produce medical glass equipment in the Rio area.

Clients and the Uses of Science

While students and staff were being recruited and trained, another very important aspect of the institute was receiving attention. The growth of a "client" relationship with national, state, and private agencies helped ensure continued financial support for the institute's work. Since the results of research are frequently unpredictable, a program of applied science activities was necessary to ensure a demand for the institute's services and products. The support of the national government was especially important, since changes in government threaten the stability of scientific organizations. The client relationship therefore had to be consolidated in such a way that perceptions about the value of the institute would carry over from government to government. At the same time the demand for applied science products could not be allowed to force the institute exclusively into the field of practical hygiene, nor destroy its scientific autonomy.

The basis for a client relationship lay in the belief that the institute, as a government agency, could provide practical solutions to problems in public health. Such a role was clearly envisaged by Cruz in his first proposal to Congress in 1903, and was codified by law in 1907–1908. A client relationship was established through a series of scientific surveys and investigations, and by the creation of temporary and semi-permanent scientific stations in selected areas of the country.[19] A few examples illustrate how these problem-oriented investigations arose and the effect they had on the survival of the institute. In 1906, for example, the privately owned Central Railroad Company requested through the Ministry of Public Works that the institute examine the source of malaria infection among its construction workers building railway lines in the state of Minas Gerais. A team of medical scientists, led by Chagas, established itself in Xerem to scrutinize charges that the Company had failed to protect its workers from infection. At the time the mortality rate among those contracting malaria was extremely high, and this prevented the Company from hiring workers, despite the offer of high wages.[20] Chagas started a program of compulsory "quininization" among workers, the first of its kind in Brazil and one which effectively reduced mortality. Artur Neiva took charge of a small isolation hospital, established at the camp site according to Cruz' specifications, enforced the quinine program, and began to study malaria-transmitting mosquitoes in the area.[21] As a result of the success of this campaign against malaria, a number of other anti-malaria programs followed. In a chapter which he contributed to a book on malaria by the malaria specialist, Ronald Ross, Cruz listed several campaigns carried out under the auspices of the institute. These included campaigns during the construction of the Itatinga Railroad, Santos, in the state of São Paulo, during the work of damming the rivers Xerem and Mantiqueira for the increase in water supplies to the city of Rio de Janeiro, at the time of the elongation of the railway lines of the Brazilian and Northern Lines of the Minas Gerais Railroad Company, during survey work for the Bahia and Espírito Santo Railway, during construction of the North West Lines of the British Railroad, and in connection with the yellow fever campaign in the suburbs of Rio de Janeiro.[22]

Yellow fever was another disease for which the Institute provided much needed technical assistance. In 1910, for instance, authorities in the city of Belém, near the mouth of the Amazon, asked the institute to organize a wholesale attack on yellow fever which, in the period 1904 to 1910, had caused the deaths of nearly two hundred people each year. Cruz went to Belém with a scientific team, hired ninety-three aides, and by means of many thousand separate inspections and cleanings, eradicated the *Aedes* mosquito from the city in six months.[23]

Two other important practical sanitation surveys of a broader kind deserve mention. The first occurred in 1910, when Cruz was requested by the Madeira-Mamoré Railway Company to survey the health conditions along the Madeira and Mamoré rivers in the Amazon in preparation for the construction of new railway lines. The expedition resulted in one of the first modern reports on the sanitary conditions of the Amazon.[24] A similar and more comprehensive report by Cruz appeared in 1913 as a result of a survey of the Amazon for the Ministry of Agriculture.[25] Unfortunately the government failed to act on the recommendations for improving public health in the area.

The significance of the practical scientific work of the Oswaldo Cruz Institute was complex. Many public works and construction of privately owned communications networks (such as railroads) were aided by preventive sanitation measures put into effect as a result of the findings of the institute. Its technicians were at the disposal of a variety of agencies to supervise offensive and defensive action against disease. In turn, the field work had the effect of extending the reputation of the institute to areas outside Rio de Janeiro, in a sense nationalizing its medical research work in Brazil. In 1906, as has already been mentioned, a small laboratory affiliated to the Oswaldo Cruz Institute was started in Belo Horizonte, later to become the Ezequiel Dias Institute. In 1917 a similar affiliated institute was organized in Maranhão. The first great national expedition of exploration in Brazil in modern times, under General Rondon, also was indebted to the institute, which trained a number of the expedition members in microbiology before their departure for the Amazon.

The extension of the microbiological sciences to new geographical areas in Brazil was made possible in part by the nature of the sciences themselves, which possessed a strong practical component. Given the shortage of funds for science and the necessarily pragmatic view of politicians and legislators concerning the allocation of resources, applied science resulted in a broadening of the Oswaldo Cruz

Institute's constituency.[26] Practical missions also promoted the internal development of the institute. In a country with little tradition in science and somewhat isolated from international communities of science, there was some difficulty in establishing the confidence essential to compete in the world scientific marketplace and produce original scientific work. The missions carried out by scientists from the institute provided an excellent mechanism for training, and for confirming the validity of training techniques by success in the field. It must be remembered that most of the scientists were extremely young at this period in the history of the institute, and lacked an older generation of experienced microbiologists to guide them. Practical missions served the purpose of building confidence and maintaining morale. Artur Neiva and Belisário Penna's long journey on horseback through the states of Goiás, Pernambuco, and Bahia in 1913 can be seen as a classic confrontation between the medical technician and the diseased population of Brazil.[27] The realization that Brazil was a "vast hospital" added to the growing awareness of the geographical, mental, and medical distance separating the privileged classes of the cities and the mass of people in the rural—and especially the northern—areas of the country. The Oswaldo Cruz Institute helped to bring about an eventual collaboration between the Brazilian government and the Rockefeller Foundation in the 1920s in order to eradicate some of the chief diseases endemic and epidemic in Brazil, such as yellow fever and hookworm disease.[28] The exposure of the actual state of public health in areas outside Rio resulted therefore in an increased awareness of the need for the institute's work.

Institutes tied too closely to practical science, however, run the risk of atrophy. Such institutions fail to attract the best students, sink into routine, and have difficulty adapting to new circumstances caused by the eruption of new health problems. Without research, institutions cannot participate actively in the international community of science. They risk dependence upon other communities of science for the fundamental research ideas that lie behind practical applications in science.

Research Science at the Institute

The development of research was therefore the third key to the vitality of the institute over a twenty-year period and more. To describe in detail the research work of the staff at the Oswaldo Cruz Institute in the seventeen-year period between 1903 and 1920 is beyond the scope of this book.[29] Some indication of the range of scientific research can be gathered by scanning the articles appearing in the institute's journal, the *Memorias do Instituto Oswaldo Cruz*, which began publication in 1909. An estimate of the growth of publications and the number of scientists at the institute in the period 1900–1919 can be made from Fig. 1.[30] The publications include those in foreign journals. In its research, barriers between basic and applied science broke down; many investigations originally undertaken for their scientific value yielded unexpected practical results, while practical studies often led to new research. There was as a result a continual and beneficial feedback from both ends of the "research and development" spectrum.

Bacteriology, the field that had led to the founding of the institute, continued to occupy a central role between 1900 and 1920. Vasconcellos, Cardoso Fontes, and Gomes de Faria were all primarily bacteriologists. The bacilli of plague, leprosy, and tuberculosis all received attention. Cardoso Fontes, in particular, made the study of tuberculosis and Koch's tubercule his specialty, his work being honored at the First International Congress of Tuberculosis.[31] Animal diseases spread by bacteria were also examined by Godoy and Gomes de Faria. At the practical end, numerous vaccines were produced, and serological studies undertaken.

Yellow fever provided another area of concentration. The fact that this was a viral rather than a bacteriological disease was not known at the time. In the 1920s, after the period of Cruz' directorship, institute scientists, like many elsewhere in the world, were caught up in the enthusiasm over the Japanese bacteriologist Noguchi's supposed discovery of the etiological agent of yellow fever, the *Leptospira icteroides*. Several expeditions by scientists from the Oswaldo Cruz Institute to Bahia, one of the major foci of yellow fever, failed to confirm Noguchi's findings, though some attributed this failure to deficiencies in the work of the Brazilian bacteriologists rather than deficiencies in Noguchi's science. But by 1928, Olympia da Fonseca had repudiated Noguchi's findings.[32] On the more positive side, the pathologist Henrique da Rocha Lima identified the pathological lesions of the liver found in yellow fever victims while still a member of the Oswaldo Cruz Institute, although the results were not published until 1911 when Rocha Lima had already left Brazil for Germany.[33]

FIG. 1 Number of staff researchers at Instituto Oswaldo Cruz and number of articles published by staff researchers, 1900-29

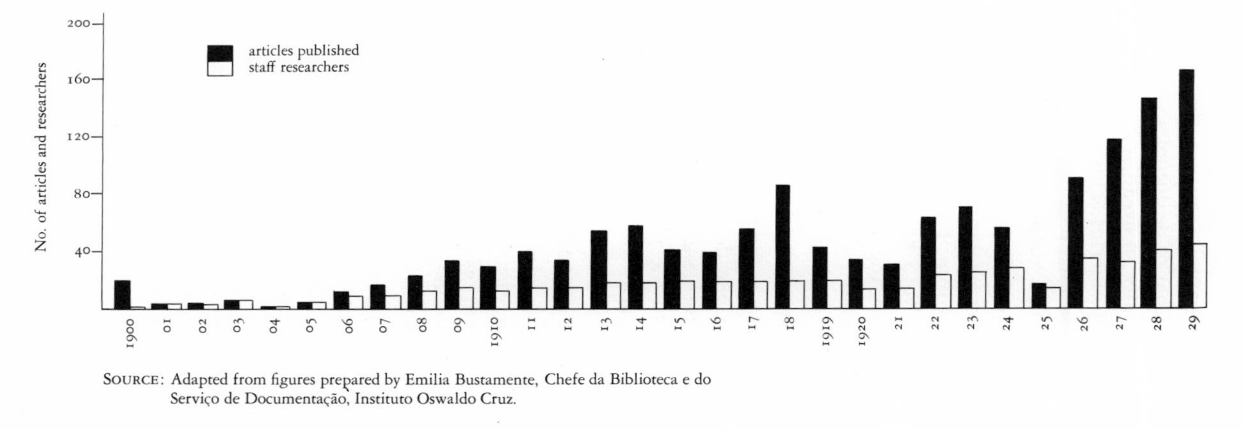

SOURCE: Adapted from figures prepared by Emilia Bustamente, Chefe da Biblioteca e do Serviço de Documentação, Instituto Oswaldo Cruz.

A third, and eventually one of the most important, fields of scientific inquiry was that of protozoology. Activity and discoveries in this area were eventually so great, we can even speak of a school of protozoology in South America. This "school" was initiated by Aragão and Chagas, and was the first field of science to attract foreign scientists to the institute, such as Prowazek and Max Hartmann.

The original interest in protozoology grew out of the very practical anti-malaria campaigns initiated for the first time in Brazil by the Oswaldo Cruz Institute. From these practical concerns grew new research activities. In 1905, for instance, Carlos Chagas was responsible for discovering that infection by the mosquito occurs primarily inside the home, a concept with immediate implications for prophylaxis. Neiva's study of malaria-carrying mosquitoes resulted in the isolation of new species and in the discovery that some plasmodia could become resistant to quinine.[34]

Meanwhile, Henrique de Beaurepaire Aragão, who had come to the Oswaldo Cruz Institute as a student to prepare his medical thesis, turned his attention to the plasmodium in the pigeon, *Haemoproteus columbae*. In 1907, Aragão announced his unraveling of the life-cycle of the plasmodium, the first announcement from the institute to generate attention in international scientific circles.[35] Aragão's work was important for the later understanding of the exo-erythrocytic cycle of the malaria plasmodium. Aragão also studied free-living protozoa; to this field Aragão attracted other scientists at the Institute, such as Aristides da Cunha, who began to study freshwater plasmodia. Da Cunha, together with José Gomes de Faria, also pioneered marine biology in Brazil.

The greatest stimulus to protozoology came, of course, when Chagas discovered American sleeping sickness (*Trypanosomiasis americana*) in 1908. This discovery was an excellent example of biological research resulting in a discovery of practical importance. The setting for Chagas' discovery was an anti-malaria campaign initiated by the institute in Minas Gerais. While in Minas, Chagas began a study of insects in general. He became particularly interested in a local biting insect, a Reduviid bug known as the "barbeiro" (a triatoma), which lived in the walls of the local dwellings.[36] In the gut of one specimen he noticed the presence of a trypanosome. Suspecting this might be disease-producing, and that human beings might be the natural host, Chagas' managed to infect marmoset monkeys through the bite of the bug, and then proceeded to discover the presence of the trypanosome in the heart and brain tissues of patients whose diverse clinical symptoms, including swelling of facial tissues and heart disease, had escaped classification until that time. The discovery, announced in Germany and Brazil, initiated years of research at the institute, with the result that Chagas' disease has become one of the most closely studied of all human diseases.[37] As a result of this discovery, Chagas was awarded the Schaudinn prize for protozoology in 1913.

Chagas' work stimulated both practical and general research. At the general level, the cycle of the causative agent, the *Trypanosoma cruzi* (named by Chagas after his friend and colleague, Oswaldo Cruz) was studied intensively, along with the cycle of other trypanosomes.[38] In 1909 Gaspar Vianna demonstrated the leishmania phase of the trypanosome, thus distinguishing it from other trypanosomes.[39] The study of the insect vector, the triatoma, led to a general examination of other triatomas, some of which also transmit the trypansome. More practically, the clinical symptoms, the incidence, the pathology, and the geographical range of the disease was subjected to much scrutiny. Originally Chagas had associated Chagas' disease with cardiac and neurological symptoms, including cretinism, and had asserted that its incidence was widespread. In 1913, the same year that Chagas received the Schaudinn prize, Chagas' colleague at the Oswaldo Cruz Institute, Artur Neiva, traveled through the rural state of Goiás and estimated that in many of the villages fully one half of the local population was either crippled by cardiac disease, paralyzed, or made imbecilic as a result of Chagas' disease. In the first years after Chagas' announcement of his discovery, however, rather few cases of Chagas' disease were seen, and by the 1920s the incidence, clinical symptoms, and even the identity of Chagas' disease was called into question. In 1922, a noisy debate between Chagas' supporters and detracters took place in the National Academy of Medicine of Rio de Janeiro. The commission appointed to make an inquiry fully confirmed the originality of Chagas' discovery and the reality of American sleeping sickness.[40] While knowledge of the clinical and pathological features of the disease has grown, a World Health Organization report on Chagas' disease in 1960 concluded that 35 million people are exposed to the risk of infection by *T. cruzi* in Latin America, while seven million people at present are in fact infected by the parasite.[41]

A spin-off from the interest in diseases involving insect vectors was the development of entomology at the institute under Adolfo Lutz.[42] Olympio da Fonseca later made the study of tropical fungi and mycological diseases part of the research program. Helminthology was added, as well as biochemistry, and, after Cruz' death, a department of physiology.[43]

"National" Science and "International Science"

Several factors seem to have been responsible for the rapid and successful development of a research program at the Oswaldo Cruz Institute. First, the capacity for carrying out research is in part a function of having a sufficiently large supply of scientists in the country, since only a certain percentage of scientists trained at any one time will actually make original discoveries. A developing country is usually hampered in its research by the small total number of scientists, but this disadvantage can be partially overcome if, instead of trying to cover several very different areas of research, a country concentrates its efforts in a few specific areas. The Oswaldo Cruz Institute possessed the advantage that there were virtually no other institutions in Brazil competing with it for scientists. Though at a later stage in the evolution of a scientific community, competition among different institutions tends to act as as stimulus to research and prevents inbreeding, at the early stage of science in Brazil the institute's ability to corner the market aided in the very rapid development of a research team.

In specializing in the field of microbiology, the institute benefited from the stage of development reached by the field by the early 1900s. Many of the fundamental ideas of bacteriology and serum therapy, as well as the concept of the insect vector, had already been worked out in some detail by European scientists by the time a team of microbiologists became organized in Brazil. At this point, Brazilians began to investigate along lines which were already established. Chagas' disease, for example, was discovered in part through Chagas' knowledge of the transmission of malaria. He reasoned by analogy with malaria that the triatoma might be an insect vector for a trypanosome that he suspected might have pathological effects in humans. Though his work was therefore closely dependent on concepts worked out with reference to other diseases, his discovery was nonetheless original and extremely important for stimulating a wide range of research activities over a number of years. Similarly, Aragão's work on the plasmodium in the pigeon grew out of efforts being made at the time to unravel the cycle of the malaria plasmodium in humans. It was much less likely that, had microbiology been in its infancy in 1900, the Brazilian scientific community would have possessed the intellectual and professional resources to contribute fruitfully to the development of the field. An interesting example comes to mind here from the United States. A. Hunter Dupree, in his study of science in the government, describes the failure of American scientists to solve the question of the transmission of Texas cattle fever in the 1870s, when the idea of the insect vector was imperfectly understood. The relative immaturity of organized science in the period compared unfavorably with the maturity and professionalization of science by 1893, when Theobold Smith successfully identified the tick as the vector in Texas cattle fever.[44]

A second reason why the microbiological sciences were especially suited to give impetus to research in a society that was still only semi-scientific was that their applied and pure aspects could not meaningfully be separated. Since they were not solely laboratory sciences, opportunities for research were continually being offered during the carrying out of practical field missions. For instance, it was during a field program at Xerem on behalf of the Central Railroad Company in 1906 that Artur Neiva discovered species of plasmodium resistant to quinine, not only adding to new techniques of prophylaxis but also providing insights into the method by which immunity is acquired. The microbiological sciences might be contrasted with other, more basic fields of science, such as physiology, where the practical outcomes of basic research are less easy to predict. Given the fact that at the time of the founding of the Oswaldo Cruz Institute, no national policy for science existed and medical research could not expect the automatic support of the government, it appears that sciences in which the gap between basic and applied aspects are great are less suited for *initiating* new phases in the institutional status of science in a developing country. Successful institutionalization of physiology had to wait for the new period of growth of the Oswaldo Cruz Institute in 1918 after Cruz' death; thus it depended upon a prior establishment of support for research science, support which could be widened to include new research areas such as physiology.

In this respect it is interesting to note that research at the institute concentrated on those aspects of microbiology that related most closely to Brazilian problems. Yellow fever and malaria were both public health risks in large areas of the country; entomological studies concentrated on Brazilian insects, and protozoology on Brazilian protozoa. A large part of the institute's research dealt with what was a Brazilian discovery and largely a Latin American disease, American sleeping sickness. In concentrating on Brazilian aspects of microbiology, of course, the institute served the most pressing needs of the country. The institute also had a comparative advantage in these areas owing to accessibility of clinical and experimental materials. Their work developed self-referentially rather than with exclusive reference to European or North American science, and allowed Brazilian scientists to become exporters of ideas in science rather than only importers. The research at the institute developed, therefore, a distinctly "national" or Brazilian character.[45] While this might be construed as a parochial limitation on the Institute's research efforts to problems with low priority among scientists outside Brazil, and a consequence of insufficient manpower and intellectual resources to tackle a wider range of research topics, it is also subject to a different interpretation. The microbiological sciences were applied sciences to which the research ideal had been attached. This process is described by Ben-David as occurring in the last third of the nineteenth century, at a time when research had a high value. In bacteriology, research tended to be directed toward specific disease problems rather than by internal developments in one of the basic sciences.[46] The orientation toward specific diseases meant that many of the subjects selected for research by Brazilians, such as the role of insects in the transmission of disease, though chosen originally for their relevance to Brazilian health problems, were nonetheless subjects of interest to scientists outside Brazil. A concentration on Brazilian problems did not rule out the possibility of making discoveries that shed light on disease mechanisms in general, or on similar diseases in other countries. Such concentration also increased the probability that Brazilian medical problems would become interesting to medical scientists outside Brazil. The problem-oriented character of the microbiological sciences explains why this field, rather than physics or physiology, was the first to receive support in Brazil. In physics, subjects for research are often dictated by the internal development of the field itself, and may only be remotely connected with the actual needs of a developing country.[47] The institutional development of the physical sciences may therefore depend upon a much higher level of support for basic research than existed in Brazil in the 1900s, since the relevance of research in physics to the country would have been less immediately apparent and therefore more difficult to justify. For developing countries, in fact, advances in the physical sciences may entail an increased dependence upon the more advanced countries of the world, both for the definition and choice of research topics, and for foreign scientific and technical assistance. In comparison, the medical sciences allow developing countries to carry out good work even when support for basic research is low.[48]

The ability of the institute to contribute to research had several important consequences for the long-term endurance of the institute. Research gave the institute flexibility in responding to new crises in public health, and in turn, in initiating new areas of research. Through publication in research journals, Brazilian scientists established contacts abroad. The international recognition this gave the institute in turn improved the institute's position in Brazil and encouraged the support of successive governments. It also prevented a migration of Brazilian scientists, who without some guarantee of a career in research in Brazil, might well have been tempted to make their careers abroad.[49] At the same time, the practical and applied work of the institute allowed those who wished to make a contribution to Brazilian development an opportunity to do so. It is remarkable, in fact, that scientists of the stature of Carlos Chagas and Adolfo Lutz carried out nearly their entire careers in Brazil.

What, in effect, had occurred at the Oswaldo Cruz Institute from 1903 on was a steady widening of its scientific capabilities, starting in areas most closely related to practical hygiene tasks, such as yellow fever eradication, and expanding over the years to include the study of protozoa, insects, worms, fungi, bacteria, and, finally, general physiology. As new scientists were recruited to the institute, new scientific interests developed and eventually new departments were organized around these interests. For example, the research of Olympio da Fonseca into mycology led eventually to the establishment of a department in this field. The program of expansion did not follow a pattern determined *a priori*, but instead built upon existing strengths and around the best scientists available. While the research program was from the beginning somewhat specialized, administratively, therefore, the institute was characterized by its diversity. The institute finally emerged by about 1912 as a school of tropical medicine, incorporating the sciences of bacteriology, parasitology, protozoology, medical entomology, and medical mycology. In this sense its founding paralleled the founding of schools of tropical medicine elsewhere in the world in the same period.

A result of the internal vitality of the Oswaldo Cruz Institute was its survival of Cruz' premature death in 1917. The forceful leadership of an institution by one man, especially a man of great political influence, can threaten the institution with collapse when the leader dies and the constituency disappears. Other problems arise when younger men are prevented from rising to positions of prominence, or when no methods are evolved for transferring the personal prestige of the leader to the institution itself. Early in his career as a man of public affairs, Cruz had relinquished his role of researcher for that of scientific administrator, though he kept on top of the research field and was first-rate at suggesting research topics to students and staff workers. His forced absence from the institute during the public health campaign had led to the delegation of authority to handle the routine tasks of administration. The post of director was in fact rotated among the department heads, and the positions of department heads among the research staff, who held the job for six months before handing it on to someone else.[50] A certain capacity for administering science and dealing with government officials was developed that was somewhat independent of Cruz himself. Within the institute, the traditional concept of the single professor dominating the work of the junior assistants broke down and was replaced by collaborative teamwork among a group of men who, spiritually at least, saw themselves as innovators.

Cruz had been increasingly incapacitated by renal disease from as early as 1908, and this had prepared his colleagues for his early retirement or death (Cruz in fact retired a few months before his death in 1917). By this time, the indicated inheritor for the post as director was Dr. Chagas, whose discovery of Chagas' disease had brought him considerable international fame. Two years after taking charge, Chagas was in fact able to almost double the institute's budget and add several new departments to the institute.[51] A few years earlier, while attending an international conference at Dresden, where the institute had again won honors for its scientific work, Cruz, normally reserved about his victories, had exclaimed that the position of the institute in the world of science was completely guaranteed.[52] Despite certain limitations in the scope of the institute's work, this claim was certainly borne out for many years.

Conclusion

It remains in this chapter briefly to place the history of the Oswaldo Cruz Institute in the wider context of changes in Brazilian social, intellectual, and economic life. In doing so, the meaning of "successful" science in a developing country is made more clear.

According to a recent history of Brazil, the process of modernization began sometime after 1850, though at first the pace of change was extremely slow.[53] Before this date the technological innovations that had characterized industrialization in Europe had not become a part of Brazilian life. After 1850, however, Brazil "was decisively swept into the vortex of the international economy."[54] There was an increased reliance on modern transport and steam and electric power, a rise in commercial and industrial activity, and the emergence of new urban groups. Traditional society slowly came under attack. By 1914, even if "the promise of modern change was unfulfilled," nonetheless new expectations had been born about the place of technology, industry, and science in the country's future growth.[55]

The question remains as to what part the Oswaldo Cruz Institute played in bringing about these changes in Brazil. During his own lifetime, Cruz was appreciated as a powerful political figure and innovator, and the institute a novel contribution to Brazilian life. Long after Cruz' death, Fernando de Azevedo, in his classic study of Brazilian culture, assessed the significance of the institute for Brazil in the following terms:

> Great as were the services Cruz rendered to Brazil . . . conquering the plague, yellow fever and malaria, they were not superior in their scientific value to the work which he undertook in nationalizing experimental medicine . . . and creating with the foundation of the Institute of Manguinhos not only the greatest scientific center of research in the country but a whole brilliant school of scholars and experimenters in the various branches of science cultivated in that institution.[56]

The implication is that the institute, far from being a unique event in the history of Brazil, with few consequences for the further development of science, acted as a catalyst for change.

There are three areas where the institute's influence was felt. The first concerns the supply of scientific manpower within the country. The second concerns government interest in sanitation. The third concerns the problem of the efficient utilization of foreign scientists and foreign technical aid.[57]

Not surprisingly, it was the industrially advanced state of São Paulo which benefited most from scientists trained at the Oswaldo Cruz Institute. The period of the 1890s had been a period in which the sanitary services of the state had been organized along modern lines. No institution could compare with the Oswaldo Cruz Institute in the scope and quality of its work, however, when in 1912 a medical school was finally started in São Paulo, some twenty years after the idea had first been discussed.[58] In 1918, Dr. Artur Neiva, long a member of the Oswaldo Cruz Institute, was invited to leave Argentina, where he had been aiding the development of a department of medical biology at the Bacteriological Institute of Buenos Aires, to take over the direction of the state's sanitary services. Neiva was responsible for drawing up a sanitary code similar to the federal code instituted by Cruz in 1903, and for reorganizing a number of the existing scientific institutions. The Butantã Institute briefly entered a new phase of productivity, began publication of its own journal in 1918 and a new program of research and training. A scientific attack on the coffee disease, *broca de café*, which was devastating valuable coffee crops, was also started. In 1927, Neiva's interest in the agricultural problems of the state led to the founding of the important Biological Institute, which was closely modeled on the Oswaldo Cruz Institute. Here it was proposed to carry out research into the agricultural sciences much as the Oswaldo Cruz Institute carried out research into the sanitary sciences. The vice-director and later director of the Biological Institute was Dr. Henrique da Rocha Lima, an early staff member at Manguinhos, who had returned to Brazil after a long and successful research career in Germany.[59]

The Oswaldo Cruz Institute also played an important role in extending the range of federal sanitary services. In 1918 Carlos Chagas was placed in charge of a federal program against the terrible epidemic of influenza in Rio de Janeiro, and in 1919 was appointed director of public health, thus linking once more the Oswaldo Cruz Institute and the national public health program. Under Chagas the institute expanded, while the sanitation program was extended into the rural areas, where a number of sanitation stations were established.[60] In 1919, ten states were involved in the attack on hookworm disease, and thirty-two sanitation posts were maintained at government expense.[61]

One reason for the extension of the federal sanitation programs was the help received from the Rockefeller Foundation. In 1913, the Foundation had set up its International Health Board to extend sanitation programs around the world.[62] Its programs were designed to strengthen local expertise. Aid was offered only to those inviting it, on the understanding that the recipient country would share the costs of programs and would eventually assume full responsibility for them if the programs were successful.[63] In 1919, under Chagas' supervision, programs against hookworm diseases and yellow fever were begun. Cooperation between Brazil and the United States increased in 1928 when yellow fever returned in epidemic proportions, owing to the relaxing of vigilance against the mosquito, and the rise of jungle yellow fever, transmitted by a species of mosquito (*A. gambiae*) distinct from *Aedes aegypti*. A number of technicians and scientists worked in close contact with Americans, with the result that a new generation of sanitation scientists came into being in Brazil. A number of researchers from the Oswaldo Cruz Institute were also sent on Rockefeller Foundation grants for further training in the United States and Europe.[64] The Brazilian government's ability to benefit from foreign technical aid in this period can be seen at least in part as a function of the existence of an adequate center of sanitation science in the form of the Oswaldo Cruz Institute, rather than merely as a form of dependency on outside help.

On the debit side in the history of the Oswaldo Cruz Institute, we should cite first its eventual isolation from university biology. In its early years the institute pioneered the teaching of microbiology for medical students. As universities were founded in Brazil, first in Rio de Janeiro in 1920 and, more importantly, in São Paulo in 1934 (considered the first modern university), the teaching of biology was taken over by the university and this separation of the Oswaldo Cruz Institute from a university structure resulted in a narrowing of its research tasks and tied its scientists more closely to applied science.

The solidarity of its research staff, which had prevented a

premature fragmentation of its research effort, and its concentration in a specialized area of science, which had given the institute its success, began to create a degree of inbreeding and fossilization. Chagas' directorship was followed by directorship of others from the original staff, such as Aragão and Cardoso Fontes, suggesting too great a reliance on a small group within the institute. The institute's financial autonomy was hurt under the Vargas regime, when reorganization led to the interruption of the sale of its vaccines and serums for profit and use by the institute, and the subordination of the institute to the Ministry of Health. The institute has suffered from political intervention during the Vargas regime in the 1930s and 1940s and, more recently, under the present military government. As other fields of science and technology have come to compete for government funds, the entire health field has suffered a relative decline in its position of importance compared to other areas, and this too has hurt the Oswaldo Cruz Institute.

Despite these setbacks, however, the tradition of biomedical research established in the almost thirty-five year span between 1900 and 1934 under Oswaldo Cruz and Carlos Chagas at the Oswaldo Cruz Institute, was, I am convinced, largely responsible for the fact that to this day biomedical research is one of the strongest fields of science in Brazil.

References

[1]These sources have already been referred to earlier in this study as the *Administrative Records*: books containing materials on personnel, terms of employment, and activities. Another extremely important source of information has been biographies of scientists where available, and various histories of the institute written by members of the institute. Of these, the most useful are: Henrique de Beaurepaire Aragão, *Oswaldo Cruz e a escola de Manguinhos.* Conferência realizada no Centro Acadêmico Oswaldo Cruz de São Paulo em 20 de septembro de 1940. Segunda edição (Rio de Janeiro: Imprensa Nacional, 1945), and his *Notícia histórica sôbre a fundação do Instituto Oswaldo Cruz* (*Instituto de Manguinhos*) (Rio de Janeiro: Serviço Gráfico do Instituto Brasileiro de Geografia e Estatística, 1950); Ezequiel Caetano Dias, *Traços biográficos de Oswaldo Cruz* (Rio de Janeiro: Imprensa Nacional, 1945), and his *O Instituto Oswaldo Cruz: resumo histórico 1899-1918* (Rio de Janeiro: Manguinhos, 1918).

[2]Henrique de Beaurepaire Aragão, *Oswaldo Cruz e a escola de Manguinhos,* p. 13.

[3]Henrique de Beaurepaire Aragão, *Notícia histórica sôbre a fundação do Instituto Oswaldo Cruz*, p. 12.

[4]This was Henrique de Beaurepaire Aragão, "Sôbre o cyclo evolutivo do halterídio do

pombo (Nota preliminar)," *Brasil-Médico* 21 (1907), 141-142, 301-303. Aragão's scientific career is described in Renato Clark Bacellar, *Brazil's Contributions to Tropical Medicine and Malaria: Personalities and Institutions.* Translated by Anita Farquhar (Rio de Janeiro: Gráfica Olympica Editôra, 1963), pp. 185-188. Between 1900 and 1907 fifteen medical theses were prepared at the institute. See Instituto Oswaldo Cruz, *Lista cronolbjica das publicações do Instituto Oswaldo Cruz de 1900 à 1915* (Rio de Janeiro: Manguinhos, n.d.).

[5]The new positions created by the bill of 1904, which created the sanitary services of the Rio municipality and the federal government and established the sanitary brigades, are itemized in Brazil, Directoria Geral de Saúde Pública, Placido Barboso e Cassio Barboso de Rezende, *Os serviços de saúde pública no Brasil, especialmente na cidade do Rio de Janeiro de 1808 à 1907* (*Esboço histbrico e legislação*), 2 vols. (Rio de Janeiro: Imprensa Nacional, 1909), Vol. 2, pp. 892-898.

[6]The budget and staff of the Bacteriological Laboratory established by the bill in 1904 is given in Brazil, Directoria Geral de Saúde Pública, *Os serviços de saúde pública no Brasil,* Vol. 2, pp. 896-897.

[7]Alcides Godoy's biography is found in Octavio de Magalhães, "Alcides Godoy," *Memorias do Instituto Oswaldo Cruz* 49 (1951), 1-6. For his work on cattle disease, see Alcides Godoy, "Nova vacina contra o carbúnculo sintomático," *Memorias do Instituto Oswaldo Cruz* 2 (1910), 11-21.

[8]Details on Rocha Lima's career from the *Administrative Records*; from Renato Clark Bacellar, *op. cit.*, pp. 180-184, and from Henrique da Rocha Lima, "Com Oswaldo Cruz em Manguinhos," *Ciência e Cultura* Vol. 4, No. 1 and 2, 15-32.

[9]Chagas' early career at Manguinhos is taken from *Administrative Records*, and from Henrique de Beaurepaire Aragão, "Carlos Chagas, diretor do Manguinhos," *Memorias do Instituto Oswaldo Cruz* 51 (1953), 1-10, and Carlos Chagas Filho, "Carlos Chagas, 1879-1934," *O Hospital* 54 (Julho 1958), 9-15. A large number of articles on Carlos Chagas are listed in Instituto Oswaldo Cruz, *Cinqüentenário da descoberta da doença de Chagas. Carlos Chagas (1879-1934), Bio-bibliografia* (Rio de Janeiro: Instituto Oswaldo Cruz, Biblioteca, 1959), pp. 9-12. This publication also lists Chagas' original publications. Chagas is another outstanding scientist from Brazil who deserves a biography.

[10]*Administrative Records.*

[11]See especially the two articles by Henrique de Beaurepaire Aragão already cited.

[12]Instituto Oswaldo Cruz, *Oswaldo Cruz Files, File No. 7A. Programa do curso de aplicação do Instituto Oswaldo Cruz.* In 1913, lectures were given by, among others, Vasconcellos, Godoy, Vianna, Fontes, Chagas, and Neiva, all of whom had entered the institute as young men.

[13]See Renato Clark Bacellar, *op. cit.*, pp. 199-201.

[14]Renato Clark Bacellar, *op. cit.*, pp. 202-209. Vianna's scientific work has been republished in Edgard de Cerqueira Falcão, *Gaspar de Oliveira Vianna 1895-1914. Opera omnia* (São Paulo: A Gráfica de 'Revista das Tribunais', 1962).

[15]See reference 42 of this chapter, and Chapter 7, for further details on Lutz' scientific career.

[16]Instituto Oswaldo Cruz, *Museum Document, Biblioteca do Instituto Oswaldo Cruz. Balanço dado em janeiro de 1911.*

[17]Ruth Martins de Magalhães, *A biblioteca do Instituto Oswaldo Cruz.* Lecture given at a conference held in the Department of Pathology, Oswaldo Cruz Institute, 18 May 1966. Mimeographed, p. 2.

[18]The microscope had in fact been used in Brazil from about the 1830s on by scientists brought to Brazil by the Princess Leopoldina. Later, in the 1850s, the Bahian school of medical scientists, starting with Otto Wucherer in his classic studies of ancilostomiasis and filariasis, employed microscopes. In Rio de Janeiro, Dr. Pedro Severiano de Magalhães, who had come from the Bahian center, also used a microscope in his study of filariasis in man. In the 1880s and 1890s the microscope was of course used routinely by those working in microbiology, but their numbers were few and their work unoriginal, as mentioned already. Fajarda, at Rio Medical School, studied the malaria plasmodium, and Lacerda studied yellow fever, but few microscopists earned a reputation for their discoveries. The best of the scientists using the microscope in the 1890s was unquestionably Adolfo Lutz, whose career and work at the Bacteriological Institute in São Paulo is described in Chapter 7 of this book. Professor Olympio da Fonseca Filho, of Rio de Janeiro, describes the history of the microscope in Brazil in his article *Primórdios da microscópia óptica no Brasil: alguns pioneiros.* Unpublished manuscript, no date. Professor Olympio da Fonseca argues that a new epoch in the use of microscopy developed with the founding of the Oswaldo Cruz Institute.

[19]See Olympio da Fonseca Filho, "Manguinhos no interior do Brasil. Filiais, postos, expedições," in Edgard de Cerqueira Falcão, *Oswaldo Cruz: monumenta histórica*, 3 vols. (São Paulo: Brasiliensia Documenta 6, 1971-73), Vol. 2, pp. 17-24.

[20]See Carlos Chagas, "Prophylaxia do impaludismo," *Revista Médica de São Paulo* 11 (1908), 391-399, and "Prophylaxia do impaludismo," *Brasil-Médico* 20 (1906), 315-317, 337-340, 419-422, and *Brasil-Médico* 21 (1907), 151-154.

[21]Arthur Neiva, "Profilaxia da malaria e trabalhos de engenheira: notas, comentários, recordações," *Revista do Clube de Engenharia* 6 (1940), 60-75.

[22]Oswaldo Gonçalves Cruz, "Prophylaxis of Malaria in Central and Southern Brazil," in Ronald Ross, *The Prevention of Malaria* (London: John Murray, 1910), pp. 390-399.

[23]Instituto Oswaldo Cruz, *Oswaldo Cruz Files, File Nos. 5 and 5A* contain the details of the campaign and the original reports written by Cruz and his colleagues.

[24]Oswaldo Gonçalves Cruz, *Considerações gerais sôbre as condições sanitárias do Rio Madeira* (Rio de Janeiro: Papelaria Americana, 1910).

[25]Oswaldo Gonçalves Cruz, *Relatório sôbre as condições médico-sanitárias do Valle do Amazonas* (Rio de Janeiro: Jornal do Commercio, 1913).

[26]Among the clients for scientists from the Oswaldo Cruz Institute, working in the field of malaria studies and anti-malaria campaigns alone, we could include the Ministry of Works Against Drought, the Service for the Defense of Rubber, railway companies, port authorities, etc.

[27]Arthur Neiva e Belisário Penna, "Viajem científica pela norte da Bahia, sudoeste de Pernambuco, sul de Piauhí, e de norte á sul de Goiáz," *Memorias do Instituto Oswaldo Cruz* 8 (1916), 74–224. The section by Olympia da Fonseca Filho, "Manguinhos no interior do Brasil" (reference 19) lists several of the other scientific expeditions undertaken by staff members of the Oswaldo Cruz Institute.

[28]See Chapter 7.

[29]Recently, in commemoration of Cruz' birth, Professor Olympia da Fonseca Filho prepared a very useful summary of the institute's scientific work from its foundation to 1972, as viewed by a member of the institute from the perspective of the present. See Olympio da Fonseca Filho, "A escola de Manguinhos: contribuição para o estudo do desenvolvimento da medicina experimental no Brasil," in Edgard de Cerqueira Falcão, *Oswaldo Cruz: monumenta histórica*, esp. pp. 24–128.

[30]I wish to thank Senhora Emilia Bustamente for her kind permission to use material collected by her in preparing this figure.

[31]For Cardosa Fontes' work see Renato Clark Bacellar, *op. cit.*, pp. 177–179. See also J. Vieira Filho, "Antônio Cardosa Fontes e sua obra," *Jornal do Comércio*, 25 de dezembro, 1932.

[32]See Olympio da Fonseca Filho, "A Escola de Manguinhos," (reference 29), pp. 34–37, for an account of Noguchi's influence on yellow fever research in Brazil in the 1920s.

[33]*Ibid.*, p. 32.

[34]Arthur Neiva, "Profilaxia de malaria e trabalhos de engenheira: notas, comentários, recordações," *op. cit.*, and his "Formação de raça do hematozoario do impaludismo rezistente à quinina," *Memorias do Instituto Oswaldo Cruz* 2 (1910), 131–140. Neiva's scientific career and contributions to entomology are described in Renato Clark Bacellar, *op. cit.*, pp. 189–198. See also César Pinto, "Arthur Neiva: cientista e homem público," *Revista Médica-Cirurgia Brasileira* 40 (1932), 2–10, and T. Borgmeier, "Artur Neiva: a propósito do seu sexagesimo anniversário natálico," *Revista de Entomologia* 2 (junho 1940), (Volume commemorativo do 60° anniversário natálico de Artur Neiva). Many of his scientific papers appeared in the volumes of the *Memorias do Instituto Oswaldo Cruz.*

[35]Henrique de Beaurepaire Aragão, *Sôbre o cyclo evolutivo do halterídio do pombo (1ª e 2ª Notas)* (Rio de Janeiro: Besnard Frères, 1907).

[36]The bug was then known as *Conorrinhus megistus.*

[37]The bibliography on Chagas in Portuguese is vast. His discovery was published in Carlos Chagas, "Neue Trypanosomen. Vorläufige mitteilung," *Archiv für Schiffs-und-Tropenhygiene* 13 (11) (1909), 120–122, and "Nova tripanozomiaze humana. Estudos sôbre a morfolojia e o ciclo evolutivo do Schistzotrypanum cruzi, n. gen., n. sp., ajente etiolójico de nova entidade mórbida do homem," *Memorias do Instituto Oswaldo Cruz* 1 (1909), 159–218. See also Instituto Oswaldo Cruz, *Cinqüentenário da descoberta da doença de Chagas*, and Carlos Chagas, "Descoberta da Tripanozoma cruzi e verificação da Tripanosomiase americana. Retrospecto histórico," *Memorias do Instituto Oswaldo Cruz* 5 (1922), 67–76.

[38]Chagas for a while created a new genus, *Schizotrypanum cruzi,* for the parasitic flagellate, so that this name appears in the literature of the period.

[39]Gaspar de Oliveira Vianna, "Contribuição para o estudo da anatomia patolójica da 'Molestia de Carlos Chagas,' (Esquizotripanoze humana ou tireoidite parazitária)," *Memorias do Instituto Oswaldo Cruz* 3 (1911), 276–294.

[40]See Olympio da Fonseca Filho, "A Escola de Manguinhos," reference 29, pp. 64–66.

[41]World Health Organization, Technical Report Series, No. 202, *Chagas' Disease. Report of a Study Group* (Geneva: World Health Organization, 1960), p. 4, and World Health Organization, Technical Report Series, No. 411, *Comparative Studies of American and African Trypanosomiasis. Report of a WHO Scientific Group* (Geneva: World Health Organization, 1969), p. 6.

[42]Lutz' scientific work is analyzed in the following articles: Carlos Chagas, "Adolpho Lutz," *Memorias do Instituto Oswaldo Cruz* 18 (1925), i–xxii; Artur Neiva, "Necrologia do Dr. Adolfo Lutz (1855–1940)," *Memorias do Instituto Oswaldo Cruz* 36 (1941), i–xxiii. A bibliography of Lutz' scientific work is found in the 1941 June issue of the *Memorias do Instituto Oswaldo Cruz.*

[43]See Olympia da Fonseca Filho, "A Escola de Manguinhos," reference 29, pp. 75–100, 103–105, 119, 120.

[44]A. Hunter Dupree, *Science in the Federal Government: A History of Policies and Activities to 1940* (Cambridge, Massachusetts: The Belknap Press of Harvard University Press, 1957), pp. 154–155.

[45]Louis Couty, the French biologist who spent several years in Brazil in the 1870s and early 1880s, argued that the best way for Brazil to participate in the general advance of science was by concentrating on the study of diseases and problems that occurred in Brazil. See Louis Couty, "Os estudos experimentaes no Brasil," *Revista Brasileira,* Primeiro anno (1884), 215–239.

[46]Joseph Ben-David, *The Scientist's Role in Society: A Comparative Study* (Englewood Cliffs, New Jersey: Prentice-Hall, Inc., Foundations of Modern Sociology Series, 1971), pp. 142–143.

[47]Some aspects of the problem of the relevancy of research in developing countries receive attention in Claire Nader, "Technical Experts in Development," in Claire Nader and A. B. Zahlan (*eds.*), *Science and Technology in Developing Countries* (Cambridge: Cambridge University Press, 1969), pp. 447-462.

[48]See Chapter 8, section *The Meaning of "National Science"*, for a further discussion of this problem.

[49]There was, instead, a degree of *internal* migration of scientists from the Oswaldo Cruz Institute to other institutions in Brazil.

[50]See Henrique de Beaurepaire Aragão, "Carlos Chagas, diretor de Manguinhos," *Memorias do Instituto Oswaldo Cruz* 51 (1953), 1, for a description of this rotation and its purpose.

[51]See Henrique de Beaurepaire Aragão, "Carlos Chagas, diretor de Manguinhos," *op. cit.,* and the budget for the Institute, 1919, in Brazil, *Leis dos Estados Unidos do Brasil,* Leis do orçamento de receita e despesa, 1919 and 1920.

[52]E. Sales Guerra, *Osvaldo Cruz* (Rio de Janeiro: Casa Editôra Vecchi Limitada, 1940), p. 563.

[53]Richard Graham, *Britain and the Onset of Modernization in Brazil 1850–1914* (Cambridge: Cambridge University Press, 1968), p. 23.

[54]*Ibid.*, p. 24.

[55]*Ibid.*, p. 319.

[56]Fernando de Azevedo, *Brazilian Culture: An Introduction to the Study of Culture in Brazil.* Translated by W. H. Crawford (New York: The Macmillan Company, 1950), p. 259.

[57]An important task for historians and sociologists of science is to establish criteria for assessing the significance of new institutions.

[58]Ernesto de Souza Campos, *Institutições culturais e de educação superior no Brasil, resumo histórico* (Rio de Janeiro: Imprensa Nacional, 1941), pp. 149–151.

[59]*Ibid.*, pp. 179–180, 597, 515–517.

[60]Henrique de Beaurepaire Aragão, "Carlos Chagas, diretor de Manguinhos," *op. cit.*

[61]For an account of the Rockefeller Foundation's collaboration with Brazil, see the *Annual Reports* between 1919 and 1925.

[62]The history of the Rockefeller Foundation and its work in the developing countries is found in Robert Shaplen, *Toward the Wellbeing of Mankind: Fifty Years of the Rockefeller Foundation.* Foreword by J. George Harrar. Edited by Arthur Bernon Tourtellot (Garden City, New York: Doubleday, 1964); and Raymond Blaine Fosdick, *The Story of the Rockefeller Foundation* (New York: Harper, 1952).

[63]The Rockefeller Foundation, *Annual Report*, 1919, pp. 23–24.

[64]For a complete list of Rockefeller Fellows from Brazil see The Rockefeller Foundation, *Directory of Fellowship Awards, 1917–1950* (New York, N.Y.), Tables 11a and 11b. The pathologist Dr. Bowman C. Crowell of the Rockefeller Foundation came to the Oswaldo Cruz Institute to head the department of pathology. In São Paulo, the Foundation cooperated in the improvement in teaching of the hygiene sciences in the new medical school, and in 1919 the Foundation organized a new Hygiene Institution in conjunction with the medical school.

7

The Bacteriological Institute of São Paulo, 1892–1914: The Role of Applied Science

It could be argued that what we see in the founding and development of the Oswaldo Cruz Institute in Brazil was merely one aspect of a worldwide phenomenon, namely the establishment of research institutions in microbiology and tropical medicine in response to a succession of medical discoveries, beginning with Louis Pasteur's work in bacteriology. Medicine had been placed on a new basis by the knowledge that many illnesses are caused by microorganisms, by the identification of the causative bacteria of many of the major epidemic diseases, and by the use of vaccine and serum therapy. The understanding of the role of the insect vector in spreading malaria and yellow fever were other important additions to medical knowledge. With these advances there came a shift away from purely clinical medicine to laboratory medicine, and the founding of a number of laboratories for research scientists and specialists. The Pasteur Institute of Paris was of course one of the most prominent and successful. The founding of schools of tropical medicine in Liverpool and London and the creation of Johns Hopkins University Medical School in the United States were other examples of this change in medical organization.

While this interpretation of the history of the Oswaldo Cruz Institute contains a certain truth, it is misleading in its blurring of the important differences that exist between the developed countries and the developing countries with respect to scientific change. A major argument of this study has been that, in a developing country, the success of a given institution of science depends upon the solving of a series of political, administrative, educational, and research problems peculiar to countries with limited resources and supports for science. If in fact it were true that the development of the microbiological sciences themselves, taken together with a degree of population and economic growth, were sufficient to ensure the survival of research institutions, or even of institutions of applied science, then one should expect to find other institutions of microbiology being founded in the same period in Brazil and enjoying the same degree of success as that of the Oswaldo Cruz Institute. On the whole, however, this is not true. A number of institutions of science were founded in the period which were to claim their share of notice from Brazilian historians, and to become prominent at a later date, such as the Butantã Institute, founded in 1899 in São Paulo. Few of these institutions became centers of research until very recently, however, and even the best were institutions of applied science which enjoyed an early success and then declined into stagnation.

A study of one of these institutions places in perspective the great importance of scientific entrepreneurship and aggressive institution-building that maintains a balance between basic and applied science.

The institute in Brazil most closely paralleling the Oswaldo Cruz Institute in its functions and the period of its founding was the Bacteriological Institute, under the direction of the parasitologist, protozoologist, and entomologist Adolfo Lutz, whose name has been mentioned several times in connection with the Oswaldo Cruz Institute. The Bacteriological Institute was founded in 1892, at a time when the state of São Paulo was experiencing rapid economic and population growth, when new supports for science and engineering were emerging, and consequently a degree of professionalization of science was taking place. It was also a period of institutional change, when many new institutions of education were being created, such as the Polytechnical School, started in 1893. The Bacteriological Institute played an important role in shaping the development of a new public health program in the state in the 1890s and early 1900s, one which made the state one of the leaders in this field in the Union. The history of the institute is important, therefore, for adding to our knowledge about an important period in an important state. Yet, paradoxically, the institute failed to grow institutionally in ways that would ensure its survival. Its research work was that of one scientist. The applied science work of the institute also began to falter when, exasperated by administrative difficulties, Lutz left the institute to join the staff of the Oswaldo Cruz Institute in Rio precisely because an opportunity for a career in research was offered to him there. Though Lutz maintained ties with the Bacteriological Institute until 1913, the institute declined and was finally closed altogether in 1925.[1] Though a negative case of science, from an institutional point of view the history of the institute is very interesting for the questions it raises about the survival of science in a developing country.

São Paulo and Public Health

The city of São Paulo had begun to experience rapid economic growth in the 1870s. This process was accelerated by the abolition of slavery in 1888 and the end of the monarchy in 1889. The state was rapidly becoming the economic heartland of the nation, with coffee the staple crop of the state. With the freeing of the provinces from the rigid centralization of the old Empire, the state government of São Paulo emerged as a powerful force in the development of the economy. While in this period the population of Rio de Janeiro was gaining rapidly, the population of the city of São Paulo, though never matching that of Rio, was growing at an even more rapid rate. In 1893, for instance, the city of São Paulo's population was 129,409, compared to Rio de Janeiro's 522,551.[2] By 1900, however, São Paulo's population had jumped to 240,000, a growth of over 100 percent in seven years; by that year Rio's had grown to 600,000.[3] A large part of the phenomenal growth in São Paulo was the result of immigration. Even in 1893, foreigners accounted for over fifty percent of the city's population; by 1907, Italians alone outnumbered Brazilians two to one.

With the rapid rise in population came a transformation of the city. New buildings replaced those of the colonial period. There were elegant hotels and many urban improvements, including gas lighting in the streets. Industries were being founded at an unprecedented rate. Foreign languages of all kinds were heard, and foreigners were welcomed as bringing new outlooks and commercial practices that would aid state development.[4] Another change concerned the development of institutions of higher learning. Under the monarchy, São Paulo could boast only a single important institution of higher education, namely, the Law School. The Law School was necessarily the focal point of intellectual life in the city. But as the vogue of positivism grew in the 1870s, some positivists declared that science rather than more traditional fields should become the backbone of education. In 1874, a Society for Propagating Public Instruction began to offer classes to a hundred students, and by 1887 six hundred and eighty students were enrolled, and a library of 5,000 volumes organized. Literacy also rose quickly in this period, after compulsory secondary education was established in 1874.[5] The rise in literacy was aided by the arrival of substantial numbers of literate immigrants into the state.

With the formation of the Republic in 1889, responsibility for the development of educational and economic institutions passed from the federal government to the state governments. In São Paulo, changes were noticeable in the two important fields of engineering and public health. Both engineers and public health technicians became an important element in the economic transformation of the city. In engineering, for instance, opportunities for a career in civil engineering began to open up. Before 1890, engineers had had to go either to Rio de Janeiro or abroad for training, many of them to the United States. With the development of railroads in the state of São Paulo, and the construction of new roads and water works, several engineers rose to positions of prominence in the city. Engineering societies and a new Polytechnic School were founded. The career of the São Paulo engineer, A. F. de Paula Sousa, which is described by Morse in his history of the city, exemplifies this period of transition in the development of the profession of engineering.[6] Paula Sousa had been born in 1843, and after training in Switzerland returned to Brazil to organize water and sewage works. He then left Brazil for the United States to work on the Rock Island and St. Louis Railroads, returning again to Brazil to direct railroad and other engineering projects. In the 1890s he was, successively, federal minister of public works, and the first director of the São Paulo Polytechnic School.

The field of public health underwent a similar process of transformation. Before 1889, the state had been poorly served by medical and sanitary institutions, and little action had been taken against epidemic disease.[7] Lack of funds from the imperial government, and a poorly developed imperial sanitary system for the provinces, kept the public health services inadequate.[8] By 1889, the situation had become more urgent, and with the collapse of the monarchy and the assumption by the state government of responsibility for protecting the public health of the state, a new era of public health opened. In 1891, the first laws establishing a rational and comprehensive sanitary system were passed by the state legislature. These laws were consolidated in 1892, when the state was divided into sanitary districts, each with its own sanitary inspector, and a centralized Sanitary Council formed in the capital.[9] At this time a number of new scientific institutions were formed. These included a Pharmaceutical Institute, a Bacteriological Institute, a Laboratory of Chemical Analysis, and a Vaccination Institute, vaccination against smallpox having in theory been made compulsory in the state in 1891. The question of the establishment of a medical school to train new doctors was left undecided at this time, owing to differences of opinion among the members of the commission called together to discuss the organization of the school. Positivists in particular opposed creating a state-run medical school. A medical school was not in fact founded in São Paulo until 1912; until then, students from São Paulo continued to go to Rio and Bahia for their medical training, as they had done traditionally.

The creation and financing of the various institutions was justified by the elite mainly in terms of the growing population and the need to protect the commercial interests of the state, rather than in terms of humanitarianism, or because of public pressure to deal with epidemic disease. Many of the immigrants entering Santos, the port city serving São Paulo, were susceptible to yellow fever, which was endemic along the coast. São Paulo was mercifully spared epidemics of yellow fever, owing to the high altitude of the city, which prevented the *Aedes aegypti* mosquito from breeding. This immunity was one factor, believes Morse, in the city's growth. However, as immigrants came into the state, epidemics erupted in cities in the interior, such as Campinas, where a bad outbreak occurred between 1889 and 1892, and Ribeirão Prêto, which suffered an epidemic in 1903. In addition, immigrants were bringing into the state many diseases that were comparatively rare in the area—for example, cholera, scarlet fever, and typhus. Together with the plague, which appeared for the first time in 1899 in Santos, these diseases were best identified by laboratory analysis, and many of them were susceptible to control by vaccine and serum therapy. The vitality of the state government and the need to protect immigrants from disease placed the new sanitary services, and with them the Bacteriological Institute, in a favorable position for future expansion.

The Bacteriological Institute: Its Scientific Work

How in fact did these social, economic, and intellectual trends affect the development of the Bacteriological Institute of São Paulo?

What kind of scientific work did the institute undertake, how was it organized and directed, and how did its organization and direction affect the long-term status of the institute? When the Bacteriological Institute was founded in 1892 its purpose was specified as that of "the study of microbiological and bacteriological problems in general and especially those concerning the etiology of epidemics, and endemic and epizootic diseases most commonly found in the state," with the production of vaccines and serums, and with the undertaking of microscopic and laboratory examinations necessary for the elucidation of clinical diagnoses.[10] Its initial functions were therefore defined more widely than those of the Oswaldo Cruz Institute in 1903. The institute was granted an annual budget of 27,000 milreis (13,500 dollars, at 50 cents to a milreis) and a staff of two professional scientists aided by two servants. In this respect it was therefore comparable to, but smaller than, the Oswaldo Cruz Institute in 1903.

To staff the institute, São Paulo officials first looked to France for a competent bacteriologist, just as Baron Pedro Affonso was to do in 1900 to staff the Serum Therapy Institute in Rio. A French professor, Dr. Felix Le Dantec, was offered a contract to come to São Paulo, and in his letter to the Brazilian Minister in Paris stipulating the conditions of his acceptance, it is clear he intended to found in São Paulo a Pasteur Institute which would provide technical and biological courses to students. He asked the state to advise him whether a technician could be found in São Paulo to aid him at the institute, or whether he should employ one in Paris to accompany him to Brazil.[11]

Le Dantec arrived in Brazil in December of 1892, but for reasons that are not clear he left Brazil four months later. Cerqueira Lemos, whose history of the institute reproduces several important documents, believes his departure was due less to problems with the institute than to Le Dantec's desire to use his visit to Brazil as an opportunity to study yellow fever. Once materials on yellow fever had been collected, Le Dantec left the country for France, and the state government was faced with finding a replacement. A happy solution to the problem occurred in the recruitment of the Brazilian scientist Adolfo Lutz in 1893.

Adolfo Lutz had been born in 1855 to Swiss parents who had emigrated to Brazil in 1849. Following training at the medical school in Berne, he had traveled in Europe, visiting many of the medical centers there. Supposedly, he met Lister in London in 1880 and Pasteur in

Paris, and worked with Professor Unna in Hamburg. He returned to Brazil to qualify himself at the Rio Medical School in 1881 and went to work in the interior to gain practical medical experience as a clinician. His real vocation was for biological research, and he soon began to publish in foreign journals on subjects such as ancilostomiasis and amoebic hepatitis. In 1889, Lutz, who had published a paper on leprosy, was invited by Professor Unna to visit the Molukai leprosarium in Hawaii, and there he married an English nurse. He returned to Brazil in 1893 to become the interim vice-director of the Bacteriological Institute.[12] His return to Brazil was one indication of the increased opportunities for a career in science existing in the state of São Paulo at this time.

Under Lutz' direction, the Bacteriological Institute undertook a series of important scientific investigations into diseases common or epidemic in the city and the state. The Bacteriological Institute was, in fact, the first center of science in Brazil to be organized along modern laboratory lines, and its work included the first systematic application of bacteriology and parasitology to public health in Brazil. Notwithstanding the hostility of several of the medical doctors in the city, Lutz, together with Dr. Emílio Ribas, who became the director of the sanitary services of the state in 1898, were responsible for greatly improving public health in the state of São Paulo. At a time when the population was growing at an unprecedented rate, the mortality coefficient of the city of São Paulo was reduced from 30.73 in 1894, to 28.27 in 1895, 31.53 in 1896, 24.86 in 1897, 21.27 in 1898, to 18.14 in 1899.[13] Prior to 1890, the mortality coefficient had been steadily rising. The relative absence of yellow fever in the state was one factor contributing to the fact that the quality of sanitation was higher in the state of São Paulo than in the federal capital of Rio de Janeiro in the first decade of the twentieth century.

The work of the Bacteriological Institute can be divided into four major areas, excluding the routine work of blood and urine analysis, and the production of vaccines and serums.[14] The investigation by Lutz into the supposed "bacterium" of yellow fever "discovered" by the Italian medical scientist Saranelli, though interesting, has also been excluded from this account.[15] The first area of investigation concerned the outbreak of Asiatic cholera on August 13, 1893, in the Hotel for Immigrants on the outskirts of the city. Lutz carried out bacteriological analysis and confirmed the diagnosis of cholera within a day. Several members of the medical profession in São Paulo refused to accept this diagnosis, claiming that what was being observed were merely cases of diarrhea. Resistance to the diagnosis was so strong that six years later a book was published in Rio to prove that cholera had never existed in Brazil. Following this first diagnosis came the diagnosis of other epidemics of cholera in the state in 1894 and 1895. A useful demonstration of the accuracy of laboratory techniques occurred in 1894, when on Christmas Eve an outbreak of what seemed to be cholera appeared in the Hotel for Immigrants in the city. Two thousand people fell ill with vomiting and diarrhea, but all except one recovered. Lutz proved that what was involved was a case of straightforward food poisoning, but to prove his diagnosis specimens were sent to the Institute of Hygiene in Hamburg, where his conclusion was confirmed.[16]

In 1895 Lutz began to study the so-called "Paulista fevers." These fevers were commonly believed to be a result of São Paulo's climate, hence the name. Bacteriological examination by Lutz revealed there was no sign of the malaria plasmodium, and that most cases were those of true typhoid fever. The Paulista fevers were exposed as a medical fiction. Since some cases of malaria were found to have been included in the traditional classification, Lutz also turned his attention to the distribution of this disease, and concluded it was endemic only along the coast and in certain towns in the interior of the state.[17] Lutz' study of the Paulista fevers led to a clash with several of the medical doctors in the city, since his classification based on causative organisms upset traditional classifications based on clinical signs and symptoms. Lutz' opinions were debated for several months at the newly founded Medical and Surgical Society of São Paulo. Doubts were finally resolved by putting the matter to the vote, the results of which went against Lutz. The vote was published in the Society's *Bulletin* as the official verdict of the medical profession.[18] In this way were the merits of microbiology decided by the São Paulo doctors. Lutz commented caustically that a

> large part of the medical profession and the daily press of this city has shown little inclination to form an objective opinion on the medical questions of the day. Instead, they systematically oppose all progress, basing their ideas on the works of authors who are either not competent or are out of date. These factors were especially present during the discussions relating to the Paulista fevers.[19]

Lutz' work eventually, however, forced a change in the state registration of these fevers. The official statistics show a slow decline in the number of "Paulista fevers" and malaria recorded and their replacement by cases registered as "typhoid fever."[20]

The Bacteriological Institute was given a share of public notice in 1899, when plague appeared in the port of Santos, an event already dealt with in some detail in Chapter 4. Lutz was commissioned to go to Santos to diagnose the disease, and as a result the state immediately established a new scientific institution to prepare anti-plague vaccines and serums on the estate of Butantã outside the city.[21] Lutz' student at the Bacteriological Institute, Dr. Vital Brasil, was appointed to head the new institution.[22] Lutz' work in this investigation is interesting because it brought Oswaldo Cruz into indirect contact with the two men—Lutz and Ribas—most vitally concerned with organizing the sanitary sciences along modern lines in Brazil, and with a state machinery best able to provide mechanisms of sanitary control. In the light of Cruz' subsequent career as director of the federal department of public health, and of Lutz' later career at the Oswaldo Cruz Institute, the crossing of their paths in Santos was probably of some significance.

The last major investigation of the Bacteriological Institute before 1908 also bore some relation to Oswaldo Cruz' career in public health in Rio de Janeiro. This was Lutz' investigation into the causes and control of yellow fever.[23] Although yellow fever in São Paulo state was confined to the coast and to cities in the interior, São Paulo shared with Rio de Janeiro a concern that yellow fever would inhibit immigration into the state. The relative immunity of the state capital from yellow fever was one reason, believed the São Paulo physician Dr. Pereira Barreto, why São Paulo was uniquely situated to carry out studies on the origin and causes of yellow fever. In a letter to Dr. Emílio Ribas, director of the sanitary services, he observed that the fact that Santos was a focal point of yellow fever entering the state and the state capital made it possible to isolate the source of yellow fever. Pereira Barreto wrote:

> In São Paulo we find ourselves in an excellent situation for fixing the epidemiology of this disease and in so doing we can do other countries a great service. Never in any other part of the world has yellow fever been in such an isolated position, naked and ready to be studied in its different phases. Numerous foci tell us where it

arises, and every day we can watch it grow and expand. In some places it appears in such a way as to leave nothing to be desired in the way of perfect experimental conditions.[24]

At the time of writing, Emílio Ribas was already groping toward an understanding of the disease. Ribas had entered the state sanitary services in 1895 as a sanitary inspector in the city of Campinas, which was undergoing an epidemic of yellow fever. Ribas noticed that recent immigrants to the city housed in the Isolation Hospital were particularly susceptible to infection, but that the children housed with them often did not catch the disease even though they were in close contact with infected adults. This suggested that yellow fever was not directly contagious, a fact proven experimentally early in the nineteenth century, yet nonetheless still heatedly debated by doctors. He also noticed there were frequent repetitions of the disease even after patients were rigorously isolated.

As a result of these observations, Ribas concluded that poor sanitation rather than direct contagion was responsible for transmission, and he determined to undertake a general program of sanitation and public works in the city, with the cooperation of the resident physicians. Streets and sewers were cleared, pools of fresh and stagnant waters dried, and unsanitary buildings condemned. Since some of these measures resulted in a destruction of the breeding places of mosquitoes, yellow fever gradually disappeared, although the exact reasons for its disappearance were still not understood.[25]

Partly as a result of his success in Campinas, Ribas was appointed director of the sanitary services of the state in 1898. In this position he had begun to work in cooperation with Lutz to find a solution to yellow fever, when the results of the Reed Commission in Havana were published in 1900. Ribas immediately published an announcement about the role of the mosquito in yellow fever in the official paper of the state.[26] He also began to orient his sanitation programs along the lines indicated by the Reed Commission. In the town of Sorocaba, for instance, Ribas began to attack the foci of mosquitoes in 1901. In Ribeirão Prêto, another city in the interior of the state of about the same size as Sorocaba, the entire public health program in 1903 was based on the Finlay doctrine, i.e., elimination of the *Aedes* mosquito. In early 1903, 810 cases of yellow fever had been reported. With the extermination of the foci of mosquitoes, the disease disappeared and in 1904 no cases were registered.[27] The work of the São Paulo doctors in the state sanitary department in Sorocaba and Ribeirão Prêto predated Cruz' own work in the public health department in Rio in controlling yellow fever by the systematic extermination of the *Aedes* mosquito between 1903 and 1907.

Lutz' work on yellow fever came to a climax in 1903 with the experimental inoculations of human beings by infected mosquitoes in the Isolation Hospital of São Paulo. Permission to repeat the experiments of the Reed Commission had been obtained from the governor of São Paulo state, Conselheiro Rodrigues Alves, as early as 1901. Because of the danger of the experiments, Lutz had laid down stringent conditions for their performance, and this delayed investigation until 1903. Confirmation in Brazil that the *Aedes* mosquito was the sole transmittor of yellow fever was an added aid to Cruz as he began his public health campaign in Rio de Janeiro that same year.[28]

Research, Students, and Applied Science in the Bacteriological Institute

From this brief description of the work of the Bacteriological Institute in the field of public health, it is clear that its services to the state were of great significance. Yet the value of its contributions to sanitation should not conceal the fact that the period of genuinely innovative work of the institute was short. It lasted from 1893 to about 1903, ending with the work on yellow fever. After this date, the public health services of the state continued to be of high quality, but the Bacteriological Institute began to decline. Its most original work was associated with Adolfo Lutz, and, as the institute became increasingly burdened with routine applied science tasks, Lutz grew impatient with the limits on his time for research. By about 1900 he was already chafing under the restrictions of inadequate budgets, poor facilities, the heavy work load, the high turnover of staff—that is, all the burdens of scientific administration in a culture with little knowledge of the needs of scientific research. In 1908, Lutz left São Paulo for Rio de Janeiro, when the Oswaldo Cruz Institute was established permanently as a center of experimental medicine. There Lutz enjoyed comparative freedom from administrative duties and a long and extremely productive career in medical research. He remained nominal director of the Bacteriological Institute until 1913, when he formally left the institute. That year the Bacteriological Institute enjoyed a brief renaissance when the German medical scientist, Dr. Martin Ficker, the former teacher of Rocha Lima, agreed to come to São Paulo for two years to advise the state government on the reorganization of the institute.[29] There was little time to put his suggestions into practice, as Ficker left Brazil in late 1914 at the outbreak of World War I. By 1925 the Bacteriological Institute was considered to be no longer carrying out useful work and was officially closed, and its functions transferred to the rejuvenated Butantã Institute. Only in 1931, in a period of institutional expansion and revitalization of sciences in the state, was the institute resurrected under the new name of the Adolfo Lutz Institute.

The reasons why the Bacteriological Institute failed to transform itself into a center of experimental medicine are complex. But first it might be asked whether it mattered? Why not be satisfied with the fact that, for a critical period between 1893 and roughly 1908, the institute undertook successfully important works in public hygiene? Could Brazil afford more than one institution of research in a particular field, when the costs were great and the returns sometimes uncertain? Did the Oswaldo Cruz Institute perhaps breed a scientific elite that was isolated from the real tasks of development in Brazil? In what way should the Oswaldo Cruz Institute serve as a model for others in Brazil?

These questions were raised and to a degree answered in Chapter 6. There it was argued that, for an institution to survive in a developing country, research, application, training, and servicing cannot be separated. When we examine the history of the Bacteriological Institute in relation to the different areas of institution-building considered in Chapter 6—the building of a client relationship with agencies who use the science produced by the institute, the recruitment and training of students, and the development of research—we find that failures in the last two of these areas, and the failure to integrate all three into a system of science, prevented the Bacteriological Institute from surviving even as a practical institute of applied science. The institute's history, as an example of what might be called "the applied science trap", is most instructive when considering different models for scientific institution-building in developing countries.

Let us take the question of clients for science first. From the beginning, the function of the agencies making up the sanitary services of the state were entirely practical. Support depended on the effectiveness of science in providing solutions to practical health problems and the political visibility of these solutions. The importance of the control of disease for the development of the state was recognized, much as President Rodrigues Alves had recognized the necessity of a major public health program in Rio de Janeiro in 1903 to improve the international status of his country. The São Paulo agencies represented the first scientific organizations in the state, and a number of talented scientists were invited to take over their direction.

On the other hand, the fact that the institute was part of a comprehensive sanitation program tended in the long run to diminish the political visibility of the Bacteriological Institute itself. The smallness of the public health problems in the city of São Paulo compared to those in Rio de Janeiro, especially the absence of yellow fever, reduced the "demonstration effect" of the sanitary sciences, since much of the work of the institute took place outside the state capital, in rural towns out of reach of the major newspapers. The work of the institute was not as widely recognized as it might have been had the institute been created to deal with a specific epidemic in the state capital, especially one threatening the stability of the state government. In addition, the fact that the work of the sanitary department was distributed among several institutions in São Paulo, of which the Bacteriological Institute was only one, may have tended to make the sanitary services as a whole better known than the Bacteriological Institute as a single unit. Here Cruz' position was very different. Cruz' political prominence in a major sanitation program, the notoriety of the yellow fever campaign, and his success in the campaign, resulted in giving legitimacy to the Serum Therapy Institute. The "demonstration effect" was heightened by the fact that he directed the institution that was seen as the chief supplier of technical expertise on which the success of the public health campaign depended. As a result of his political prominence, Cruz later demanded and received important increases in the institute's autonomy and budget. In comparison, the Bacteriological Institute's budget made only small increases between 1893 and 1904, from 45,000 milreis to 50,400 milreis, at a time when the value of the milreis was decreasing. Between 1902 and 1904 the institute's budget was even cut back slightly, owing to a reduction in the budget of the public health department. The Bacteriological

Institute also continued to be a dependency of the public health department, and lacked the freedom to expand on its own.[30]

As stated earlier, crises of major importance, such as the threat of epidemics of bubonic plague, may encourage the creation of new laboratories. However, crises in public health do not necessarily produce conditions favorable for the later expansion of such laboratories. Without aggressive institution-building, institutions are often unable to survive the period of crisis and to become permanently established on a broader basis. The Butantã Institute, founded in 1899 in São Paulo in response to the threat of bubonic plague, is a good example; in its early years it lacked adequate funds and buildings, and its work in snake serums, though of considerable importance, was not wide enough in scope to attract students in large numbers. It was only in 1918, during the reorganization of the sanitary services of the state under Artur Neiva, that the Butantã Institute was revitalized and made into a modern center of science.[31]

The history of the Oswaldo Cruz Institute showed that a large part of the institute's success was due not merely to the solving of a major crisis in public health, but to the recruitment and training of students, so that a supply of researchers was ensured in Brazil. The Bacteriological Institute was much less successful in this second area of institution-building. By the law of 1892, the institute was given a professional staff of two medical doctors, plus two lower-level technical posts. In 1896, when the sanitary services were reorganized, the staff amounted to one director and three medical assistants. Many times, however, the staff posts were not filled. Of the original staff of 1894, consisting of Artur Vieira de Mendonça, Cariolano Barreto Burgos, and J. Roxo, the latter died and Barreto Burgos left shortly after. In 1897, Lutz' most important student, Vital Brasil Mineiro da Campanha, joined the Bacteriological Institute, only to leave in 1900 to take over the Butantã Institute, founded to produce anti-plague vaccines and serums. At this time Lutz was assisted by Dr. Artur Vieira de Mendonça and Dr. José Martins Bonilha de Toledo. Mendonça left in 1900, at the same time as Vital Brasil, following a dispute with Lutz over the role of the mosquito in yellow fever, and his place was taken by Carlos Luis Meyer, who remained for some time, eventually becoming interim director when Lutz departed for Rio de Janeiro in 1908.[32]

The staff was therefore never bigger than three or four medical doctors. One reason for this may have been the relatively small budget of the institute, and the fact that the institute never enjoyed a period of independence from official scrutiny, as occurred in the Oswaldo Cruz Institute when Cruz was director of public health. Another possible explanation for the lack of students training under Lutz was the absence of a medical school in São Paulo until 1912, which reduced the source of students in the area who might have been attracted to a career in the microbiological sciences.[33] It also meant that the young doctors working with Lutz found it difficult to obtain clinical experience with which to broaden their laboratory work. Oswaldo Cruz, for instance, drew heavily on medical students from the Rio Medical School to supply the staff of his institute. Yet it is also true that earlier scientists located in Rio de Janeiro, such as the physiologist, João Batista de Lacerda, working at the Physiological Laboratory in the National Museum, had not been successful in calling upon this fund of talent. More important, therefore, than the mere existence of the Medical School in Rio, was the fact that Cruz was an effective teacher, and offered students an opportunity to acquire new skills through training programs not offered elsewhere. In addition, the lack of a medical school in São Paulo was partly compensated for by the vitality of the sanitary services themselves, which provided places of employment for a number of young Paulista doctors seeking a career in medicine in the state. Dr. Emílio Ribas, for example, first began his scientific career in the state as an employee of the sanitary department in 1895. He was a pioneer in sanitation and Lutz' most important ally between 1893 and 1908 (Ribas became director of the sanitary services in 1898). Another recruit to microbiology was Dr. Vicente de Carvalho, the first director of the Vaccination Institute organized in 1891, and later the first director of the new medical school founded in São Paulo in 1912. Vital Brasil, Lutz' colleague until 1900, had also come to São Paulo in 1898, after graduating from the Rio Medical School, to make his career in the Bacteriological Institute. The economic expansion of the state and the vitality of intellectual life compensated to a certain extent for the lack of a medical school in São Paulo, so that Lutz' failure to attract new recruits to the Bacteriological Institute must be attributed in part to his failures as a scientific administrator rather than merely to the lack of a supply of medical students.

Here Lutz contrasted with Oswaldo Cruz. Oswaldo Cruz was an

extremely successful teacher, who made the training of students central to the institute. Moreover, Cruz, possibly realizing the role of the administration in a developing country, withdrew from original investigations to devote himself to the institutional problems of science in Brazil. Adolfo Lutz, on the other hand, was a man of enormous intellect, an expert linguist (he was the offical translator for the scientific papers published in the *Memorias do Instituto Oswaldo Cruz*) and was known primarily as a researcher. He was a member of a European family, and some found him cold, ironic, sarcastic, and inaccessible. Above all, he disliked the task of administration, at a time when administration had to be very important for the success of an institute. The opposition of the traditionalists in São Paulo was almost continuous throughout Lutz' stay in São Paulo, and Lutz was easily exasperated by this opposition. Once the mortality from disease had been reduced to a satisfactory figure, as it had by about 1900, Lutz' interest in running the Bacteriological Institute began to fade. According to one biographer, "As soon as the grim fight against fell disease was won, Lutz, who was by nature a pioneer and by vocation a scientist, began to weary of the daily round of dull routine."[34] He did not enjoy the challenge of establishing a center for research in an environment hostile to his plans.

The small size of the staff and the absence of students, however, made it difficult for him to escape from the "dull routine." The Institute lacked the human resources to escape from the heavy load of practical tasks demanded of it by the sanitary services, and lacked the financial resources to hire technicians to carry out routine work and free the professional scientists for research, as the Oswaldo Cruz Institute was eventually able to do. The work carried out by Lutz and his two or three assistants in a single year was very great. In 1900, for instance, Lutz examined over 7,000 rats for the plague bacillus, made 963 examinations for Koch's tuberculosis bacillus, 155 examinations of urine, 674 examinations of blood, 284 examinations for amoebic dysentery, and in addition prepared a number of vaccines and serums.[35] Lutz was also a member of forty different scientific missions between 1893 and 1906, many of them outside the city. This tended to reduce the attractiveness of the institute as a training center for students, as no other scientist of Lutz' competence was available to give instruction during his absence.

Another measure of the slow growth of the institute was the poor development of its physical facilities. In 1896, the institute moved from rented rooms in downtown São Paulo to new facilities in the Isolation Hospital in São Paulo. From the beginning, these facilities were inadequate in terms of space and equipment, and partly owing to lack of funds and partly to Lutz' unaggressive style of administration, new materials and equipment to maintain the facilities were not forthcoming. Lutz' reports contain many references to the decline in the quality of the laboratories, and by 1913 the situation was so bad that laboratory technicians were catching diseases from their own specimens, owing to inadequate measures for sterilization and poor water supplies.[36] The poor facilities and cramped space of the institute were an added disincentive to students.

Without students or an expanding staff an institution is in danger of collapsing before a sufficient concentration of scientists has developed. Dr. Vital Brasil's departure from the Bacteriological Institute in 1900 prematurely reduced the viability of the institute. In addition, an institution that is very small lacks the capacity to survive the departure of the founder. When Adolfo Lutz left the Bacteriological Institute for Rio in 1908, the research program of the institute left with him, and the institute declined further into its role as a practical laboratory of limited capabilities in sanitation. The institute could also not provide the scientific manpower to improve new institutions of science being founded at the same time, such as occurred with the Oswaldo Cruz Institute at a later date. As a consequence, the Bacteriological Institute could not influence the development of new scientific traditions. Nor could it establish contacts with foreign scientific communities that would bring foreign scientists to Brazil.

The Decline of the Institute

It was a measure of the demise of the Bacteriological Institute that a few years after Lutz' departure in 1908, the state invited Professor Martin Ficker, Henrique da Rocha Lima's former teacher in Berlin, to come to São Paulo on the occasion of the formal separation of Lutz from the institute, to undertake an evaluation of the institute, inaugurate a program of student training, and propose methods for its transformation into a modern center of bacteriology. Ficker stayed in São Paulo for almost two years, and drew up a detailed report for the state government.[37] In the light of the foregoing analysis of the Bacteriological Institute and the role of applied science institutions, his remarks on the institute are particularly interesting.

Ficker found that the institute did not satisfy any of the requirements of a modern bacteriological laboratory. The institute was too small, its library facilities inadequate, its physical conditions poor (even water supplies to the main facility and a proper chimney for carrying away noxious fumes were absent). At bottom, however, lay the failure to elevate the work of the institute above the level of the purely practical:

> Nothing is more dangerous for the development of the hygiene sciences than "schematism," that is, imitation. One cannot pretend that experiences in one country apply more or less directly to another country; on the contrary, the hygiene sciences must evolve in the locality where they are going to be applied . . . Only by understanding the outside influences on people, through knowledge of the conditions of life of the local population, can the hygienist suggest proper public health measures. The fundamental principle, therefore . . . in order for practical measures to do what they are supposed to do, and benefit the local population, is that such measures are based on genuine scientific knowledge. In Latin America there is a lack of such research work, work which embraces all areas of hygiene.[38]

Ben-David, in his sociological study of scientific productivity, also emphasizes the need for the cultivation of research on a fairly broad front, even in mission-oriented institutions. Because "the frontiers between basic and applied work are continually shifting, the establishment of specialized institutions in a field which is promising today may immobilize resources at a future date when other fields have become more interesting."[39] In 1913, Martin Ficker made a similar point. The focusing of attention on practical applications of sanitary techniques that had characterized the work of the Bacteriological Institute had resulted in an inability to handle new public health programs:

> The institute which has only practical goals ends by becoming fossilized. It can resolve some practical questions along narrow lines, and this certainly helps the country, but when the institute

is faced with new problems, it runs into difficulties because it lacks a real scientific base.

Ficker added to this judgment of the Bacteriological Institute of São Paulo the following comment on the Oswaldo Cruz Institute in Rio:

> It is one of the great services of Oswaldo Cruz that he founded for the first time in South America a medical experimental institution based upon a thorough knowledge of the natural sciences, which serves at the front line of research into infectious diseases and the application of serum therapy in a way brilliant enough to win him international prestige. It is, in Latin America at least, the only institute which is dedicated in a constant and superior way to the progress of the medical sciences . . . Already a major benefactor of Brazil for his purely practical undertaking of the combat of yellow fever, Cruz nonetheless has always placed the scientific goals of the Institute above everything else.[40]

What is interesting is that the Bacteriological Institute, during its most creative period in the 1890s under Lutz' direction, resembled in many ways the kind of institution it is sometimes believed is most suited to developing countries. Given the shortage of financial resources, trained scientists, institutional supports for research science, the cost of research, and the uncertain results from research, it is sometimes argued that institutions should be encouraged to focus primarily on applied science. Yet the history of the Bacteriological Institute shows that confining work to applied science and development of an applied science "service" orientation resulted in the stagnation of the institution. As Ficker argued in 1913, "only scientific research can open the door to practical results."[41] The history of the Bacteriological Institute also shows that escape from the "applied science trap" can only occur through aggressive recruitment of students and the providing of facilities for research, and that this depends in turn upon the skills of the scientific administrator.

The analysis of the history of the Bacteriological Institute of São Paulo, and the development of what I call "service science" shows that there is no one-to-one relation between socioeconomic development and the development of science. Although science will probably not develop institutionally in countries that are economically poor, a growing economy alone cannot guarantee the survival of institutions of science in developing countries. The key factors prove to be related to the internal administration of scientific institutions, to the legitimacy of the role of the scientist, and the construction of a balance between pure and applied science. The failure of the Bacteriological Institute to gain some measure of political visibility, and the failure to solve certain problems in the internal organization of the instituion, resulted in the Bacteriological Institute's eventual stagnation and collapse.[42]

References

[1]The institute was later re-opened, but not until 1940, when it merged with the newly created Institute Adolfo Lutz, did the work of the institute enter a new period of productivity.

[2]Brazil, *Anuário de estatística do Brasil*, 1966, p. 38.

[3]*Ibid.*

[4]Richard M. Morse, *From Community to Metropolis: A Biography of São Paulo, Brazil* (Gainesville: University of Florida Press, 1958), p. 152.

[5]*Ibid.*, p. 155. This law, however, was not enforced.

[6]*Ibid.*, p. 157.

[7]Americo R. Netto, "O caminho para a formação de serviço sanitário de São Paulo de 1597 à 1891," *Arquivos de Hygiene e Saúde Pública* 7 (Janeiro 1942), 5–34. In 1889, only 0.4 percent of the federal budget was spent on public hygiene, hospitals, public assistance, etc.

[8]See "São Paulo, Inspectoria de Hygiene, Relatório aprensentado ao Ea Inspectoria Geral de Hygiene do Imperio pelo Inspector de Hygiene da Provincia de São Paulo, Dr. Marcus de Oliveria Arruda, 1887," *Archivos de Hygiene e Saúde Pública* 1-2 (1936), 93–104. Toward the end of the Empire, Conseilheiro Rodrigues Alves helped the incipient hygiene services of the state by withdrawing money from the state budget set aside to aid immigration and placing it at the disposal of the state hygiene services. As Secretary of the Interior in 1889, he argued in his annual report that the sanitary services required more support than any other department.

[9]São Paulo, *Leis e resoluções decretadas pelo Congresso legislativo do Estado de São Paulo em 1891* (São Paulo: Diário Official, 1892). Lei n. 12, *Organiza o serviço sanitário de São Paulo*. See also São Paulo, *Actos do poder legislativo do Estado de São Paulo* (São Paulo: Diário Official, 1892), Lei n. 43, *Organiza o serviço sanitário do Estado*, pp. 25–28.

[10]São Paulo, *Collecção das leis e decretos do Estado de São Paulo de 1893* (São Paulo: Diário Official, 1893), Lei n. 240, Reorganiza o serviço sanitário do Estado de São Paulo, artigo 27, p. 153.

[11]Le Dantec's original contract and his plans for the Bacteriological Institute of São Paulo are discussed in Fernando Cerqueira Lemos, "Contribuição à história do Instituto

Bacteriológico, 1892–1940," *Revista do Instituto Adolfo Lutz* 14 (1954) (Número especial), 16–19. See also Bruno Rangel Pestana, "Cinqüentenário do Instituto Adolfo Lutz," *Revista do Instituto Adolfo Lutz* 2 (1942), 181–190.

[12]For Lutz' biography see M. Sabina de Albuquerque, "Dr. Adolfo Lutz," *Revista do Instituto Adolfo Lutz* 10 (1950), 9–30, and Artur Neiva, "Necrologia do Dr. Adolfo Lutz, 1855–1940," *Memorias do Instituto Oswaldo Cruz* 36 (1941), i–xxiii. The latter includes a bibliography of 211 publications. See also R. C. Briquet, "Adolfo Lutz: exemplo e gloria da ciência médica brasileira," *Revista do Instituto Adolfo Lutz* 1 (1941), 203–216, and *Revista do Instituto Adolfo Lutz* 15 (1955), *Número único. Número comemorativo do centenário do nascimento do Adolfo Lutz.*

[13]São Paulo, Directoria do Serviço Sanitário, *Annuário estatístico da secção demographia. Anno de 1899*, p. 79.

[14]The analysis of the work of the Bacteriological Institute of São Paulo has been gathered from several sources, the most important of which are: (1) São Paulo, Secretário de Estado dos Negocios do Interior, *Relatório apresentado ao Presidente do Estado de São Paulo pelo Secretário dos Negocios do Interior*, 1893, 1894, 1895, and 1896; (2) Adolpho Lutz, "Relatório dos trabalhos do Instituto Bacteriológico durante o anno 1897," *Revista Médica de São Paulo* 1 (1898), 175–187; (3) Adolpho Lutz, "Resumo dos trabalhos do Instituto Bacteriológico de São Paulo, 1892–1906," *Revista Médica de São Paulo* 10 (1907), 65–87; (4) Adolpho Lutz, "Trabalhos do Instituto Bacteriológico do Estado de São Paulo durante o anno de 1898," *Revista Médica de São Paulo* 2, 1899, 300–321; (5) São Paulo, Instituto Bacteriológico, *Coletanêa de trabalhos do Instituto Bacteriológico, 1895–1933*, 2 vols. Volume 2 contains Lutz' annual Relatórios for the years 1893 and 1894; (6) São Paulo, Instituto Adolfo Lutz, *Departmento de Administração*, *Relatórios* of Adolfo Lutz for the years 1895, 1896, 1899, 1900, 1901, 1902, 1903, 1906. Handwritten originals. These *Relatórios* were not published; (7) Fernando Cerqueira Lemos, *op. cit.*

[15]Saranelli had been invited to Montevideo in Uruguay in 1895 to direct a small institute of hygiene there. In 1897, Lutz and Artur Mendonça, his assistant, visited Saranelli to verify Saranelli's work, and for a while Lutz believed that Saranelli had in fact isolated the causative organism of yellow fever. See Fernando Cerqueira Lemos, *op. cit.*, 34–42.

[16]For the work on cholera, see the *Relatórios* cited above and Adolpho Lutz, "Contribuição à história da medicina no Brasil segundo os relatórios do Adolpho Lutz como director do Instituto Bacteriólogico de São Paulo (1893–1908)," *Memorias do Instituto Oswaldo Cruz* 39 (1943), 177–252. This article contains reprints of Lutz' official reports on the cholera epidemics, namely: (1) A epidemia de cólera de 1893. Relatório especial: Directoria do Instituto Bacteriológico. (2) Relatório de janeiro, 1894, fevereiro 1895. (3) Relatório de 1895. Cólera asiática. (4) Relatório de 1896. Cólera asiática e enteritis coleriformes. (5) Relatório de 1897. Cólera asiática.

[17]See the *Relatórios*, and Adolpho Lutz, "Reminiscências da febre typhoide," *Memorias do Instituto Oswaldo Cruz* 31 (1936), 851–865.

[18]L. Rezende Puech, *Sociedade de Medicina e Cirurgia de São Paulo, Memória histórica, 1895–1921* (São Paulo: Casa Garraux, 1921), and "Febres paulistas, paracer da Sociedade e conclusões finaes, 1 de 13 dezembro," *Boletim da Sociedade Médica de São Paulo* III (1897).

[19]Adolpho Lutz, "Relatório dos trabalhos do Instituto Bacteriológico durante o anno de 1897," *op. cit.*, 176.

[20]São Paulo, Instituto Bacteriológico, *Coletanêa de trabalhos do Instituto Bacteriológico, 1895-1933*, Vol. 1, *Serviço sanitário do Estado de São Paulo.* Instituto Bacteriológico, "A febre typhoide em São Paulo," por Bruno Rayel Pestana (1918), pp. 231-277.

[21]See the *Relatórios* of the Bacteriological Institute, and Instituto Adolfo Lutz, Departmento de Administração, Vital Brasil, *Relatório sôbre a peste bubônica em Santos, apresentado ao Dr. Director do Instituto Bacteriológico em 27 de novembro de 1899*, and the *Estado de São Paulo* for October and November, 1899.

[22]For the subsequent history of the Butantã Institute see Eduardo Vaz, *Vital Brasil* (São Paulo: São Paulo Editôra S. A., 1950); Eduardo Vaz, *Fundamentos da história do Instituto Butantã* (São Paulo: Instituto Butantã, 1949); Vital Brasil, *Memória histórica do Instituto de Butantan* (São Paulo: Elvino Pocai, 1941); and L. Miller de Paiva, "História científica do Instituto Butantã," *Publicações Médicas* 17 (1946), 59-72.

[23]See the *Relatórios* of the Bacteriological Institute, and Adolpho Lutz, "Reminiscências da febre amarella no Estado de São Paulo," *Memorias do Instituto Oswaldo Cruz* 24 (1930), 127-142; Emílio Ribas, "Communicações. Quinto Congresso Brasileiro da Medicina e Cirurgia, 1903. (Rio de Janeiro, 16 de junho à 2 de julho de 1903). O mosquito e a febre amarella, trabalho da Directoria do Serviço Sanitário de São Paulo," *Archivos de Hygiene e Saúde Pública* Nos. 1-2 1936), Número especial, 270-306; Emílio Ribas, "Prophylaxia da febre amarella: memória apresentada ao 5° Congresso Brasileiro da Medicina e Cirurgia," *Revista Médica de São Paulo* 6 (1903), 477-485, and 6 (1903), 504-516.

[24]São Paulo, Secretário de Saúde e Assistência Social. *Correspondência de Emílio Ribas* (setembro 1946). Collection of typewritten copies of letters sent to Ribas. Letter from Luís Pereira Barreto, 10/5/1900.

[25]Emílio Ribas, "Prophylaxia da febre amarella: memória apresentada ao 5° Congresso Brasileiro da Medicina e Cirurgia," *op. cit.*

[26]*O mosquito como agente da propagação da febre amarella* (São Paulo: Diário Official, 1901).

[27]Emílio Ribas, "Prophylaxia da febre amarella: memória apresentada ao 5° Congresso Brasileiro da Medicina e Cirurgia," *op. cit.*

[28]Emílio Ribas, "A extinção da febre amarella no Estado de São Paulo (Brasil) e na cidade do Rio de Janeiro," *Revista Médica de São Paulo* 12 (1909), 198-209; and Fernando Cerqueira Lemos, *op. cit.*

[29]See section *The Decline of the Institute* in this chapter.

[30]The official budget of the Bacteriological Institute between 1893 and 1904 was as follows: 1893, 27,000 milreis; 1894, 45,000 milreis; 1895, 45,000 milreis; 1896, 47,400 milreis; 1897, 47,400 milreis; 1898, 46,400 milreis; 1899, 50,400 milreis; 1900, 50,400 milreis; 1901, 50,400 milreis; 1902, 58,000 milreis; 1903, 43,800 milreis; 1904, 50,400 milreis. Source: São Paulo, *Collecção das leis e decretos do Estado de São Paulo de 1893*, 1894, 1895, 1896, 1897, 1898, 1899, 1900, 1901, 1902, 1903, 1904.

[31]See reference 22, this chapter, for works on the history of the Butantã Institute.

[32]See Fernando Cerqueira Lemos, *op. cit.*, for an account of staff appointments and turnover. During the whole period of Lutz' tenure of the directorship of the institute between 1893 and 1908, the professional staff never grew beyond three or four. Officially, the Oswaldo Cruz Institute was also supposed to limit itself to a staff of this size, but the actual size of the Institute was increased by the attendance of numerous students and researchers, some of whom were in fact paid staff members. In the case of the Bacteriological Institute, the medical staff was not always kept complete even during epidemics. See Adolpho Lutz, "Contribuição história da medicina no Brasil segundo os relatórios do Adolpho Lutz como director do Instituto Bacteriológico de São Paulo (1893-1908)," *op. cit.*

[33]For the history of the founding of the Medical School in São Paulo in 1912, see Pedro Dias de Silva et al., "Notas para a memória histórica da Faculdade de Medicina de São Paulo," *Annaes da Faculdade de Medicina de São Paulo* 1 (1926), 1-77.

[34]See the introduction to Adolpho Lutz, "Contribuição à história da medicina no Brasil segundo os relatórios do Adolpho Lutz como director do Instituto Bacteriológico de São Paulo (1893-1908)," *op. cit.*

[35]Adolpho Lutz, "Resumo dos trabalhos do Instituto Bacteriológico de São Paulo, 1892-1906," *op. cit.*

[36]As early as 1897, the first year in the new buildings, Lutz complained in his official report of difficulties with the gas and water supplies, and the lack of stables for horses used for serum production.

[37]Martin Ficker, *Programma para a reorganização do Instituto Bacteriológico, 1913*, unpublished manuscript obtained from the archives in São Paulo, Instituto Adolfo Lutz, Departmento de Administração. Part of the report was published in Fernando Cerqueira Lemos, *op. cit.*, 98-101.

[38]Martin Ficker, *op. cit.*

[39]Joseph Ben-David, *The Scientist's Role in Society: A Comparative Study* (Englewood Cliffs, New Jersey: Prentice-Hall, Inc., Foundations of Modern Sociology Series, 1971), pp. 161-162.

[40]Martin Ficker, *op. cit.*

[41]*Ibid.* See Chapter 8, section *National Capabilities in Research and Applied Science*, for further discussion of this point.

[42]See Fernando Cerqueira Lemos, *op. cit.*, for the history of the institute after 1914.

8

Science in a Developing Country: Some Policy Issues

Introduction

My purpose in this last chapter is to discuss some general questions about science in developing countries, questions dealt with implicitly throughout the book but which now need to be addressed directly. The discussion will lead me far from my specialty, the history of science, into debates in the science policy literature concerning the function of science and technology in development. Many developing countries are creating science policy agencies in order to integrate science and technology programs with national plans for economic development. Many difficult choices about the allocation of resources for science and technology are going to be made. Are there any guidelines for such choices? Is the science policy literature based on the industrialized nations helpful? What would "effective" science mean for a developing country?

In what follows I have concentrated on three areas which appear central to any discussion of science in developing countries, and with which my own historical study of the biomedical sciences in Brazil has been concerned. These are (1) the need for national capabilities in basic research, applied research and technology; (2) the meaning of "national" science; and (3) the institutional locus of research science.

National Capabilities in Research and Applied Science

Independence and productivity in science is a goal of almost all developing countries. Yet examples of successful research institutions are still rare. The Oswaldo Cruz Institute is one such example; let me then briefly recapitulate the nature of its "success."

The Oswaldo Cruz Institute was founded in 1900 when, as Ben-David has shown, research science emerged and changed the types of institutions and bureaucratic relations necessary for science.[1] By 1900 the pursuit of science depended on the strength of research-oriented institutions in which scientists worked cooperatively, sharing the ideal of research and passing on its methods and values to students. As we have seen, as a consequence of the rise of research science, even a partial change in Brazil's scientific output after 1900 required not merely an injection of money, but the creation of a tradition of research science completely new to Brazil. Ben-David argues that the requirements of research science were met in the United States by the new graduate universities, which were able to provide a home for research science. In Brazil, given the absence of universities and a large industrial base, the requirements of scientific research explain the wide range of institutional mechanisms associated with the survival of the Oswaldo Cruz Institute between 1900 and 1920. As discussed in Chapter 6 these mechanisms included the recruitment and training of scientists, the creation of a client relationship with the government and other agencies that could be expected to use the scientific knowledge produced by the institute, and the development of a research program.

In the development of its research, to take the last of these requirements first, the institute was not consciously shaped by a political policy demanding relevant science. Yet the responses of the institute to a series of social, political, and economic factors in fact resulted in a research program that focused on Brazilian diseases, on biological mechanisms related to such diseases, and on practical applications of such knowledge.

In its research the institute relied almost entirely upon Brazilian scientists, only some of whom had been trained abroad. Lack of incentives for foreign scientists to work in Brazil, and his own nationalism, led Cruz to rely on Brazilian personnel.[2] In a short time he was able to build up a sufficiently large group of students and physicians to form a critical mass of researchers all concentrating on related areas.

Owing to Cruz' position as director of the large federal public health program against smallpox, bubonic plague, and yellow fever begun in 1903, the demand for sanitation technicians was on the rise. The work of the institute was linked to a "market" for its product, the market being the federal sanitation program. Additional clients were found in state and municipal governments and in private companies needing the services of medical scientists. In this way the new training facilities were linked to employment opportunities, so that the small number of microbiologists trained in Brazil could begin to expect to lead professional careers within the country. In short, the success of the Oswaldo Cruz Institute between 1900 and 1930 (when politics interfered with the running of the institute) was in part the result of the creation of an interlocking *system*, involving basic and applied science, the training and employment of scientists, and the production and consumption of scientific knowledge within Brazil.

In this respect the Oswaldo Cruz Institute was most unlike earlier science efforts in Brazil. In the few instances of research work being carried out in Brazil before 1900 such work was sporadic, isolated, unconnected with any national interests, or incapable of being self-sustaining. Throughout much of Latin America today the science that exists is similarly sporadic and its component parts disjointed. Scientific research is often unconnected with the "consumption" of the knowledge it produces. For example, in the countries that make up the Andean Pact (Venezuela, Ecuador, Colombia, Bolivia, Peru, and Chile), scientific research is isolated almost completely from industrial processes. The reasons for this are complex and are discussed later in the chapter, but the consequence of the over-emphasis on university science is that over fifty percent of the regional effort in science is in the universities, while industrial manufacturing counts on only nine percent of the total scientific and technical personnel.[3] Since research is often unrelated to problems of national interest, scientific work has a tendency to become even more academic. The ideal becomes that of basic research, an activity sponsored for its contributions to new knowledge or "high" culture. Since science is not "consumed" at home by the industrial process, the only career possible for scientists is that of university teacher or researcher. But since universities can absorb back as teachers and researchers only a small percentage of the scientists they produce, the only alternative for many scientists in developing countries is to migrate, causing further loss to the country. So paradoxically, although most countries in Latin America have too few research scientists, and spend too small a percentage of their gross national product on science for science to be productive, they also produce too many research scientists to be used effectively at home.[4]

It is this chronic lack of integration of the component parts of the national science efforts of developing countries into a *system* that prevents their science from being productive and hinders the transition from stage two, dependent science to stage three, independent science. I stress the word system, involving different kinds of science, training, and employment, and production and consumption of knowledge in the success of the Oswaldo Cruz Institute in bringing about a transition in the biomedical sciences in Brazil to a semi-independent state, because studies in recent years by historians of science, sociologists, and economists indicate that the key to successful and productive science in the industrial world lies precisely in the creation of such a scientific research *system*.

This system emerged slowly.[5] In the seventeenth century the philosopher and statesman, Sir Francis Bacon, dreamed of a society in which all the creative efforts of scientists and technologists were directed toward the betterment of the human condition. Yet until the late nineteenth century science could do little to contribute to man's material life. Science could not cure his bodily ills nor increase his material output. Historians are generally agreed that the industrial revolution of the late eighteenth and early nineteenth centuries owed little to scientific theory; while scientists publicized their need for state support, little was forthcoming so long as science offered few practical rewards.[6] It was not until about a century ago that the beginning of a union occurred between science and industry. Starting in the German dye industry, and then spreading into other fields such as metallurgy and the electrical industry, science began to affect industry via practical applications.[7] Industrialists found that to remain competitive they had to keep up with the sciences in their own fields, and industrial laboratories such as the Bell Telephone Laboratory were founded.[8] Through such laboratories, and through the vigorous entrepreneurship of industrialists, the lag between scientific discoveries of potentially useful application and their actual application in industrial processes was slowly narrowed.

It was not until World War II, however, that the union between science and the state was consolidated. During the war it was realized that scientific knowledge was a form of political, economic, and military power and that science and technology could be mobilized in the service of the state. The Manhattan Project in particular showed that increased investment and mobilization in science could speed up the process of discovery as well as close the time lag between discovery and innovation. The success of war-born science in producing items such as radar and DDT led to the belief that continued investment in science would have beneficial effects on other areas of the economy and social life. As Salomon points out in one of the most recent and most important books that deal with the development of science in Europe and the United States, the priorities of political power dictated that military defense and fields of science related to it would come first in the budget.[9] But the idea of the mobilization of resources to achieve useful results took hold in all spheres, from the industrial to the medical.

How this "scientific research" system works is not entirely clear. It has proven difficult to evaluate inputs and outputs.[10] There is agreement that the connection between research and application is almost never direct but rather indirect, i.e., it is brought about through such things as institutional overlap between science and technology and by entrepreneurial attitudes toward knowledge.[11] The flow of ideas and information has traditionally been regarded as starting at the basic end of the research and development spectrum, and terminating in the application end. It is more accurate to think in terms of a two-way flow, and the success of the scientific research system depends on maintaining this flow. Techniques and technologies produced by engineers can influence scientists in their research; scientific research produces knowledge which an engineer may at times find useful for his work. Salomon describes the system as follows:

> . . . [P]ure science is merely one element among others in the *system* constituted by research activities: it no longer takes precedence on the road leading to the resolved enigmas of the universe. All contemporary research consists of reciprocal feedback between concept and application, between theory and practice, or . . . between the 'mind which works' and 'the matter which is worked.'[12]

The interdependence between science and its applications does not necessarily occur automatically. On the contrary it has to be fostered. That is to say, just because a scientific discovery has potential applications does not mean that an application will necessarily follow.[13] The actual development of an application of science depends on a great variety of factors. In one case, economic need, such as the need to remove a bottleneck in production, may stimulate such an application. In another case the existence of a market for the new application may be crucial; in yet another, the existence of technical manpower, or sufficient capital. In the late nineteenth century, it was the private industrial entrepreneur who usually was responsible for effecting greater contact between science and its applications, through the creation of industrial laboratories. In the current scientific research system of industrial nations many agencies and organizations help the system function, from universities, to government laboratories, to industrial laboratories. The whole industrial system is in a sense geared to the production of technical innovations and is the work of a very complex set of interlocking agencies and institutions.

When we turn from industrialized countries to developing countries, we find that not only is the size of the research and development system and its industrial base very much smaller, but that an integrated system of science and technology does not exist. Furthermore, recommendations about what a developing country should do in the area of science and technology, as well as existing dependency in science and technology, sometimes lead to further separation between the component parts of its science system rather than greater unification.

Let us start with the debate about basic and applied science and the implication that developing countries should not undertake basic research. Basic science is science carried out for the sake of increasing our understanding of nature rather than for the sake of any potential applications such knowledge may have. Basic science is considered to be universal in its truths, international in character, and freely available to all. It is universal because discovery is recognized according to universally accepted procedures of proof and reproducibility, it is international because all scientists participate in the scientific community, and it is freely available because science depends for its advancement on the immediate and open publication of its results.

Given the nature of research science, it is sometimes argued that a developing nation need not invest heavily in basic research within the country but should instead rely on the research work of the more advanced nations, which have greater facilities and resources for carrying it out. The idea of "national" science is repugnant to the scientist; internationalism in science means that the developing nation is not at a disadvantage in the area of basic research. Since a developing country is concerned with the potential uses of science, rather than the creation of new knowledge, it can apply the results of research carried out elsewhere. Harvey Brooks argues in *Minerva*, for example:

> Basic science, unlike technology, is truly international in its value. "Know-how" does not have to be purchased but can be freely shared among all who are intellectually prepared to appreciate and use it. In consequence the basic research in one country can be readily appropriated for application in another . . .
>
> I think that to talk about a national science policy in basic research is self-defeating and contradictory. In basic science, science policy increasingly has meaning only on a world scale.[14]

For a developing country to depend on research scientists from abroad, however, is in effect to isolate one part of the scientific research system from another. When such isolation occurs, it is an illusion to suppose that research results from outside will be "easily applicable" to problems at home. To apply research results requires, in the first place, potential users capable of understanding research and of scanning research publications for their potentialities.[15] Application may require further research to pursue new leads. Unless a developing country has its own science and development system it may well not possess enough scientists capable of scanning the research field for likely applications, much less of making such applications effectively.[16]

More important, research in the industrialized world is not pursued randomly; though its results are internationally verifiable, it is still nationally funded and the fields that tend to be emphasized are, not unnaturally, those of potential value to the funding countries, not the developing world. As a result many research fields of importance to the developing world have small priority in the developed world. It is estimated, in fact, that of all the research that is carried out in the world only about two percent is relevant to the problems of the developing world.[17] The total "capital" of international knowledge may be great, therefore, but may not include the very knowledge most needed by the developing world, knowledge which may well have to be created by the developing countries themselves. The study of yellow fever, for example, slowed in the United States after 1878 because yellow fever disappeared as a medical threat in the northern states of the Union, while the southern states were politically and financially unable to undertake their own research in the first decades after the Civil War. Since yellow fever continued to be a serious problem in Latin America throughout the period, research into the disease was carried on by Latin Americans despite poor facilities for doing so, and was only taken up by North Americans when occupation of the port of Havana in Cuba was threatened by outbreaks of yellow fever among the occupation forces in 1900.[18]

The whole argument about basic research is misplaced, because the history of the sciences in Europe and the United States shows that in the last sixty years or so distinctions between basic and applied science have been breaking down and the frontiers between them disappearing.[19] While there are some research investigations that are more general and theoretical in their conclusions than others, and other investigations that have highly specific applications as their goals, the most important aspect of science in the industrial world is that science and its applications are ultimately intertwined. The decision to call a piece of work basic rather than applied often depends more on the type of scientist carrying out the work and his place of research (university or industrial laboratory) than the character of the work itself. In making exactly this point, Salomon gives as an example the simultaneous announcement of the synthesis of an enzyme in 1969 by a university (Rockefeller University) and an industrial laboratory (Merck), one supposedly therefore carrying out basic and the other applied research.[20] In the Oswaldo Cruz Institute distinctions between basic and applied research similarly disappeared; the ability of the institute to survive as an effective institution of applied science depended on its ability to carry out a certain amount of basic research at the same time.

The arguments in favor of developing capabilities in applied science within developing countries are also many. Applied science is directed towards the solution of particular problems, such as disease, agricultural production, or the extraction of resources. While it is probable that much applied science in the industrial world has some relevance to the developing world (though it may be costly or inaccessible), much is also directed toward goals of little interest to developing nations. In the medical field, for example, most research is concerned with temperate rather than tropical diseases. As a consequence, developing countries cannot rely on industrialized countries for answers to questions about tropical illness.

Conversely, even where knowledge has been created that is useful to developing countries, local conditions may greatly affect the success or failure of its actual application. For example, high-yield varieties of corn or rice may not thrive in new soil and climate conditions; successful production may depend on careful research into local factors as well as the use of fertilizers.[21] In the extraction of resources, in agriculture, and in medicine, the need to study local conditions is clear.

Successful applied science, in consequence, demands the existence of well-equipped local teams able to assess problems *in situ.* This is turn leads back to research, since the creation of applied science teams cannot occur independently of research science. An applied scientist may find that the knowledge needed in his work demands cooperation from a researcher working in a fundamental discipline. Any institution involved in applied science must possess sufficient resources, therefore, to undertake a certain amount of research. Furthermore, applied scientists can only be trained by scientists who are themselves familiar with research techniques and who preferably are engaged themselves in research. Otherwise, training will be inadequate, as students will receive from their teachers research knowledge that will soon be out of date.

The need for constant interaction between basic and applied sciences is especially great in developing countries because applied science is often defined so narrowly as to preclude the ability to deal effectively with new problems as they arise. A good example of this is the Bacteriological Institute of São Paulo. The narrowly defined function of the Bacteriological Institute and the isolation of its applied work from research contributed to its demise as a viable institution. The Oswaldo Cruz Institute, on the other hand, slowly expanded its activities from that of a supplier and manufacturer of medical products hitherto purchased outside the country (vaccines and serums), to the invention of new vaccines and serums. Research work into yellow fever and Chagas' disease was added to its roster of activities. While specializing in the study of Brazilian diseases, this did not prevent the scientists at the Oswaldo Cruz Institute from making some contributions in science that were of interest to scientists outside Brazil. A "national" science existed that was not entirely cut off from the world of science outside, nor yet totally dependent upon that world outside for the definition of appropriate research subjects. In this way, the institute escaped from the pressure to develop a research program with high prestige but little relevance to the health needs of the country.

In the biomedical field, in fact, distinctions between pure and applied science seem particularly inappropriate.[22] This being so, the argument in favor of creating indigenous institutions in biomedical science in developing countries is especially strong. And while no developing country can probably afford to support a large arsenal of "pure research" or cover all fields of science, it is clear that, within limits, the attempt must be made to create at home an interlocking system of basic and applied research in chosen areas of scientific enquiry.

National Capabilities in Technology

This book has not dealt with the history of technology in Brazil. Current debates about technology in developing countries, however, indicate there are several factors impeding the development of independent science and technology in such countries which are not usually taken into consideration by historians of science.[23] At the risk of oversimplification I will review some of these arguments, because these arguments strengthen the case for creating national capabilities in science.

Technology, the "know how" rather than the "know why" end of the science system, has many features that distinguish it from science. The history of technology shows us that until fairly recently science and technology were independent of each other. Technical invention in the nineteenth century arose out of preceding technological knowledge rather than from science. Technologists were distinct from scientists in their education, places of work, social class, and culture. Such distinctions often remain today. Moreover, technology is also distinguished from science by the fact that technology can be owned. Technological invention is completed in the process of patenting rather than publishing, and industrial secrecy and the need to maintain competitive leads protect technical information. Patents give the right of ownership to invention. Results and discoveries in science, on the other hand, are published rapidly and openly, so that all researchers in the field can learn of them and criticize them.[24]

The differences between science and technology are such that interactions between the two are almost never direct. Indeed, much technological innovation today owes little to science but is related rather to economic factors. Nonetheless, in the industrial countries the science component of some technical innovation is high, and constant interactions between science and technology exist even though these interactions occur indirectly. Sometimes the connection is institutional; scientists in university departments serve as consultants to industry and are made aware of technical problems. Conversely, plant engineers have usually been trained in university departments of science and are therefore aware of research developments in their fields. Scientists sometimes function as engineers, and vice versa. Technical innovations may stimulate scientific advance, while scientific theory may result in technical innovation.

When we turn from the scientific system in the industrial world to that of the developing world, however, technology is not viewed as part of a system but once again as "things" which can be purchased and incorporated into the industrial process when the need arises. Technology becomes a matter of "technology transfer" and industrialization is viewed as suffering constraints only because developing countries possess insufficient capital, or inadequate markets, or an insufficient supply of trained personnel to make the technology work.

On the basis of such a view of technology it is sometimes argued that, since what developing countries need is "more" technology, the most efficient way to obtain more technology is to rely on its importation and adaptation from abroad. Since many developing countries are "late industrializing" they are in a position, it has been argued, to profit from the huge amount of technological information that already exists in the world. The wastage of re-invention will be minimized, while the purchase of technology that has been tried and tested will allow for the smooth development of industrialization.

Industrialization by the importation and adaptation of technology and technological knowledge has in fact been the usual method of industrialization in Latin America. It is because the results have been disappointing that a group of social scientists, usually referred to as "dependency theorists," have searched for the cause of this disappointment in the industrialization process itself. They argue that, in the case of countries which have been politically independent for a long time and have been heavily dependent upon foreign technology, technological dependency, far from helping the standard transfer of technology or the process of national development, has helped create an international division of labor in which Third World countries contribute labor and primary goods, and the industrialized world technical and scientific knowledge and manufactured goods.[25]

The starting point for this view of technology is the study of the history of the industrialization process in Latin America. It is generally agreed by historians that industrialization had its first burst of growth in Latin America during World War I and the depression when Latin American governments were faced with a cut-off in goods previously imported from Europe and the United States. As a result, governments encouraged the manufacturing of these previously imported goods at home. Industrial growth occurred primarily in the consumer goods industry by replicating at home the plants and products once located elsewhere. The limitations of such "import substitution industrialization" are summarized by Hirschman:

> It forecloses any fundamental adaptation of technology to the characteristics of the importing countries, such as the relative abundance of labor to capital.[26]

Reliance on foreign technology, that is, results in the use of technologies that are inappropriate to the local situation or at the very least are not designed to maximize the special mix of local factor endowment. Concentration on the consumer sector leads to a neglect of industrialization in sectors more closely related to national need, such as food production. Owing to the use of ready-made technologies, learning from technology is reduced:

> Import substitution industrialization thus brings in complex technology but without the sustained technological experimentation and concomitant training which are characteristic of the pioneer industrial countries.[27]

While in some countries import substitution industrialization is still in its early stages, in others, such as Brazil, there have been great developments in the basic industries of steel, hydroelectric power and even nuclear energy. Industrialization, however, continues to depend on the use of foreign technology. Developing countries thus find themselves paying large sums for patents, royalties, and licenses. Technology is often bought as a "package," in which the purchase of one technology entails the purchase of intermediary technologies and capital goods, each with its own costs to the country.[28]

The situation has been exacerbated in recent years by the growth of foreign ownership of industry in Latin America, especially by the rise of the multinational corporations. The effect of the multinational corporation is often not to stimulate technical innovation but rather to prevent the development of an integrated science and technology system related to the developing country's own goals. Owing to their great accumulation of technological knowledge and capital resources the multinational corporations have little need to use technology or technologists from the developing country where their subsidiaries are situated. Instead, corporations rely on technologies and patents developed in the mother plants. Research and development is carried out in laboratories in the mother countries so there is no need to contact local laboratories or institutes of research. Owing to their near monopoly of ownership of some areas of manufacturing and the ability to depend very often on their own managers and engineers, multinational corporations contribute to the "underemployment" or unemployment of scientists and technicians within host countries, even while the governments of those countries are doing their best to increase their indigenous technical manpower. The multinational corporation, by the very nature of its operations, contributes to the separation between research, university education, and application in the host country. When there are reduced employment possibilities in industry, students will not enter a degree program in the field. This in turn means that institutions in the host country cannot staff their industrial research programs. If there is no research related to industrial production, all further development is dependent upon foreign investment, which tends to perpetuate the cycle.[29]

In the United States, industrial growth entails the support of industrial research, contacts with university science, and employment for scientists and technicians. The research, development, and application components of the system continually interact in the service of producing useful, or innovative, or marketable technologies which in turn stimulate further innovations. But in developing countries where industrialization is dominated by multinational corporations, many of these backward and forward linkages never occur.[30]

The discussion concerning science in developing countries has led thus far to the following conclusions: (1) what makes science an effective part of the modern industrial system is the integration of research. applied science, and technology into a single system with a flow of ideas and information in both directions, from technology to applied science, applied science to research, research to applied science; (2) this integrated system is the result of a very large research and development effort and the involvement of the state in science; (3) it is exceedingly difficult for developing countries to develop such a system for themselves owing to the small industrial base, the ties between domestic industry and the international economy, the lack of domestic technical manpower, but above all the fragmentation of their research, development, and technological efforts. Foreign firms, especially multinational corporations, affect the growth of integration between the component parts of the scientific-industrial system by interrupting connections between such things as local technological effort and applied research and between industrial plants and employment of technicians. It is not enough, therefore, merely to increase the supply of manpower, or university-trained scientists and technologists, without finding mechanisms for integrating them into some kind of system.

Most developing countries, in fact, find themselves locked into a cycle of difficulties graphically summed up by Claire Nader as follows:

> The lack of scientifically trained leaders perpetuates inappropriate acceptance of advanced countries priorities in research and education. This dependence on others for problem definition and inappropriate education and training abroad and at home in turn produces inadequate leadership. Thus organizations and policies for science and technology are not sufficiently developed in such key areas as (a) education and training, (b) identifying and defining key research problems, (c) adapting technology to their own special situations, and (d) increasing communication between government officials and the scientific community for a greater awareness of the strength of science as an instrument of policy. The net result is that science and technology are not brought to bear fully on the development process.[31]

The Meaning of "National" Science

But to what degree can science be made to serve national goals? This question raises the issue of the possible meaning of a "national" science, and the political nature of scientific choice.

It is sometimes claimed that science cannot and should not be constrained by social, political, or economic goals. The classic expression of this position is found in Polanyi's well-known article,

"The Republic of Science," in which the scientific community is described as a well-organized, semi-autonomous republic, with its own rules and regulations for running science.[32] These include open publication, self-criticism, and control over the quality of work. Polanyi argues that for science to function properly and to make advances, it cannot be interfered with by outsiders or outside criteria such as political, social, or economic need; to constrain science only results in inferior science. What Polanyi fears is that politicians who know and understand little of the needs of science should be in a position to instruct researchers on how to go about their work: which methods to follow, which materials to choose, what results to obtain. The example of Lysenkian genetics is the kind of interference he has in mind, where ideology led to the repression of Mendelian genetics and the decline of Soviet biology.

While none would disagree that political interference with the content of science results in inferior science, it must be kept in mind that Polanyi's community of scientists refers specifically to only one part, and then only a small part, of the total science effort of a nation, namely, the basic research community. Within the total science and technological system, however, many fields of science and technology are ripe for exploitation by applied scientists and technologists. In addition, basic research is funded because such research has unpredictable but, in the long run, expected social, military, or economic returns. In this broad sense the science research system described by Salomon is already guided by criteria lying outside science. Guidance comes, for example, when funding is made in certain areas of science rather than others, since this affects the total number of scientists and technicians working in the field, and therefore the rate of exploitation of a field of knowledge.

In this respect developing countries are similar to developed ones. The politics of science dictate that science is supported for its eventual benefits to the nation. Science is funded because policy makers believe that science can help solve problems of agricultural production, water supply, or the extraction and value-added transformation of raw materials. As a consequence policy makers in developing countries believe that political and economic criteria will be used in setting priorities for their national science policies. But whereas in industrial nations basic research is funded because it "feeds" a large science and technology effort in which development and production form the largest part, in developing countries there are limited funds for science and a small supply of trained scientists and technicians. Under these circumstances the political nature of scientific choice is sometimes translated into the belief that the major effort should be in the sciences related to problems of national urgency.

Here we are not talking about political interference with the content of science, but about the use of science as an instrument of national development. But can science be manipulated in this way?

The suggestion immediately presents a dilemma because most scientists fear that, given the returns expected from science and the need for "useful" science, only applied science will be supported. They are also convinced that this is not the way to obtain solutions to specific problems. They believe there is a definite limit to the degree to which science can be manipulated, and that this limit is due to the very nature of science itself. Scientists argue that science develops at its own pace, dictated by the internal conceptual state of science, a pace which cannot be hurried up merely because of political or social need for solutions to practical problems. For example, the "need" to cure cancer is high yet the solution to the problem of cancer may have to wait for the maturation of a basic discipline in biology, such as immunology. It may not even be clear yet where, amongst the various fields of biology, advances will do most to clarify specific problems of the disease. Any expected returns from funding scientific work for cancer, therefore, are long-term, and must be assessed in relation to the internal state of the biomedical sciences.

This tension, between the need of policy makers to allocate resources to science according to political and economic criteria, and the need by scientists to be free to pursue research at their own pace, has been explored by Alvin Weinberg.[33] Weinberg suggests that in the allocation of funds for science, two rather different sets of criteria must be used by policy makers. The first set relates to the internal condition of science, the second to external criteria such as the social urgency of a problem. The internal criteria, relating to the state of science itself, are two: (1) Is the field of science ready for exploitation? (2) Are the scientists in the field generally competent? On the whole it is the scientists who must provide the answers to these questions. On the external side, Weinberg proposes three criteria: (1) Does the field have technical merit? i.e., are the technical results feasible and are the goals worthwhile technically? (2) Does it have scientific merit? By this he means does work in this field nonetheless illuminate other scientific fields close to it? (3) Does the field have social merit? This last he believes to be the most difficult to deal with since it is often impossible to judge the relevance of a piece of research for human welfare.

Weinberg's is obviously a highly rationalistic view of the problem of scientific choice. In reality, in developing countries as elsewhere, a very complicated set of political, social, military, and economic criteria are used in allocating funds, many of them beyond "rational" analysis and having to do with the history and culture of the country and the nature of the existing government. The main point is that there is no single scale of values by which criteria can be rationally set up. Weinberg does, however, point up the need of policy makers who allocate funds to science to be aware that the feasibility of projects varies according to the internal condition of science, and must be weighed against the urgency of the problem. Some scientific research involves large areas of unpredictability, so that the sciences themselves need to be allowed to mature further, at their own pace, before results can be expected. Here the best policy is to fund a wide variety of individual scientists to do basic research in fundamental disciplines. In other cases, owing to the maturity of the field and its ripeness for exploitation, rapid results can be expected by the creation of an adequate number of scientists and by the provision of sufficient resources. It is under circumstances such as the latter that additions of money and scientific manpower make all the difference between success or failure in achieving desired practical ends.

In the case of the Oswaldo Cruz Institute it was exactly the right combination of international scientific and national political circumstances that led to the successful creation of a Brazilian or "national" school of science and in the choice of fields for research. In 1900 the microbiological sciences ranked high on the criteria of readiness for exploitation. The fundamental concepts of the germ theory of disease, of immunology, and of serum therapy had been worked out in the three preceding decades and had led to immediate practical results. In 1870, in contrast, when French and German scientists were just beginning to develop knowledge of basic concepts, the field was obviously much less ripe for exploitation. I suggested in Chapter 6 that Brazilian scientists would have been less likely to be successful in effecting a transformation of science at an institutional level in 1870, given the state of the microbiological sciences at the time and the need of the institute to produce immediate results in order to continue to receive government funding.

Because the microbiological sciences were problem-oriented sciences, research at the institute concentrated on Brazilian diseases. But this did not rule out the possibility of making discoveries of interest to medical scientists outside Brazil, while it probably increased the chance that Brazilian medical problems would become interesting to medical scientists elsewhere. The state of the microbiological sciences was such that research topics considered important to international medicine did not differ greatly from topics considered important to Brazilian policy makers.

The mutually supportive relationship between national and international aspects of medical science makes medicine a particularly rewarding field of science for developing countries. Since in medicine the barriers between basic and applied science are constantly breaking down, when scientists concentrate on national medical problems they need not be isolated from the international community of medical scientists. Moreover, continued support of basic research in medicine is nearly always justified in developing countries, because discoveries in basic biomedical fields will eventually illuminate a great variety of specific, national, medical problems. In this sense it makes sense to talk of developing a national school of research.

From the policy point of view, this discussion of "national" science leads to the conclusion that scientists in developing countries must be included in the science planning process, so that reasonable expectations can be held about obtaining results from funding science, and good advice received by planners about the support levels needed for basic research. Let us suppose, for example, that the government of Brazil decided to revive the tradition of excellence in tropical medicine that had been built up in the first two decades of the century. Such a policy would clearly be justified by the high incidence of tropical diseases in the nation. The solution to some of these diseases would be found to lie at hand and to require mainly the application of large enough resources of money and manpower. In other cases research would be found to be far enough advanced within Brazil to suggest possible applications to specific diseases, but would require the support of further applied science efforts. In yet other cases there might be problems of great urgency in which the related sub-fields of science were not ready for exploitation. Scientists would have to be funded to carry out research in a wide variety of basic disciplines, and their work would overlap research in these same disciplines being carried out by biologists at the forefront of science in other countries. The Brazilian government's program for biological research could only succeed if it were guided by those Brazilian biologists who were members of the international community of science. Any possible results from this research would be uncertain and undoubtedly a long way off. What I am arguing for, once again, is the need for developing countries to create indigenous capabilities in all kinds of science.

One last point must be made concerning national science. In pursuing national goals in science a developing country may well have to carry out research into urgent problems in which, according to Weinberg's definition, the related sciences are not yet ripe for exploitation. Any number of political factors might enter into such a choice, including the overriding one: the desire of developing countries to create a science that deals with problems of national importance. It might well be that, owing to the lack of interest in a particular research field in the industrial countries, little work has been carried out in a fundamental area of science that, given special opportunities, is closely related to the solution of an urgent problem in a developing country. How would a developing country undertake work in a field that is both urgent in terms of its social or economic priorities, but not ready for scientific exploitation?

One of the few articles to examine such a problem (for developed countries) is "A Contingency Model for the Devising of Problem-Solving Research Programs: A Perspective on Diffusion Research."[34] Here the authors examine two distinct features of scientific work, what they call its "predictability" and its "urgency." Predictability is equivalent to Weinberg's "readiness for exploitation" and refers to the degree of development of the science. When a science has reached a certain point in its evolution, its further evolution can be predicted even though the exact details of this evolution are unknown. Before this point, the degree of unpredictability is greater. Urgency, which is equivalent to Weinberg's social need, refers to the intensity of the need felt for a solution to a particular problem. What kinds of scientific organizations and funding will be most useful for fields with different degrees of feasibility and urgency? On the basis of studies of group and individual problem-solving behavior, the authors conclude that groups solve problems that rank high in predictability, while individuals do best with problems low on predictability. Similarly, groups should suggest the scientific approaches to problems high on urgency and predictability, but individuals should make their own choices on research strategies when it comes to problems that have low predictability, because the possibility of group consensus on the right method is low, and such things as individual style and insight play a larger role. Situations of high urgency but low predictability are obviously those causing greatest stress, and make the greatest demands on the competence and variability of the scientific community.

This is the area where a developing country will experience the greatest difficulty, owing to the small size of its scientific community and the fact that its greatest concentration will probably be in the applied science area. Nonetheless the analysis suggests that highly urgent and unpredictable fields of science, which may have to be explored by a developing country because it cannot rely on the developed world to solve its problems, should be dealt with by the support of a scientific research community in which some scientists are involved in applied science and others in basic science. The analysis further suggests that not all science can be tied to practical ends because there is no way to know in advance which researches in the basic sciences will lead to discoveries that have potential uses. Since a degree of uncertainty is involved, some open-ended research in the basic sciences is essential if a developing country is to be able to respond flexibly to new problems as they arise.

The Setting for Science: Universities and Research Institutes

The last issue I wish to consider is that of the location of research in a developing country. Is there an ideal location? To answer this, we should first inquire what is known about the location of research in the industrial system and what types of organizations encourage productivity in science, and then ask whether conditions in the developing countries would provide a somewhat different answer to the question of the location of research.

When we look at the industrialized countries, the first thing we notice is the enormous variety of institutions connected with science—universities, research institutions, industrial laboratories, libraries, scientific societies, funding agencies, equipment manufacturers. In the developing world, we notice their lack of variety and number. It is the great variety and overlap of institutions in the industrial society that allows for the continuous flow of information between basic and applied science and technology, and for the constant monitoring of science for practical possibilities by potential users. The integration of research, application, and development in a developing country must necessarily take a very different institutional form. Owing to the paucity of scientists and institutions of science in a developing country, the organizations in the country responsible for research will have to fulfill many functions normally carried out by other institutions in the developed world. Each institution must itself provide incentives to carry out research, must bridge the gap between training and employment, must encourage further uses for research, and even supply the library facilities necessary for research. The history of the Oswaldo Cruz Institute showed that success in science depended on the institution uniting in one place research, application, training, and entrepreneurial activity, because of the lack of supporting and interlocking institutions elsewhere.

It is in this context that the debate about the role of university science needs to be restructured. In industrialized countries universities are felt to be the most important centers of research and the key to successful and productive science. Historically, the university was associated in western Europe with the very emergence of the scientists' new role in the seventeenth century. Although in the late seventeenth and throughout the eighteenth century, scientific societies rather than universities were important in maintaining interest in science, by the middle of the nineteenth century the university was well on its way to reasserting itself as the single most important center of science. The idea of "research science," in which university professors undertook original investigations and introduced students to research, was originally an innovation of the German university. The idea spread to other university systems, notably the American graduate universities founded in the 1880s and 1890s. Here the European research ideal was adapted and modified. The idea of research was broadened to include not only the sciences but other disciplines. Since the first decades of the twentieth century, the decentralized, competitive, research-oriented university has remained the single most important factor in the productivity of science.[35]

Because of the historical connection between universities and science, and because there exist almost no examples of good science in the industrial world without a good university system, it is assumed that the university will also be the home of science in the developing countries. Yet the structure of the university in many developing countries is such that the argument may need modification. In addition, the need to integrate research and application may also lead us to question the role of university science. Though the universities in some Latin American countries have an impressive history, many of them having been founded long before a single university was founded in the United States, the tradition of university research is very new. The training mechanisms found in these universities are not always favorable to the development of research. University education is extremely academic, with students being led step-by-step through undergraduate training, training in the basic disciplines of science, and being introduced to research only in the final postgraduate phase. This late introduction to research may discourage students from entering research science. The emphasis in the university on basic research as opposed to applied science may prevent the best students from considering applied science. Especially when there is an imbalance between supply and demand in science—as there often is in developing countries—students may more easily choose basic research over applied research. Yet because not all the students can be employed by the university as university teachers and researchers, the only alternative for them is to emigrate to other countries.

The isolation of the university scientist from industry, or, in the case of physiology and biology, from the health system, also creates problems, for it reduces the possibility of exchanging ideas and information between different kinds of science. The disciplinary structure of the university itself often prevents effective interdisciplinary contacts from occurring and from group approaches to research. The professor's vested interest in retaining his authority within his department may impede variety in research. Professors often have part-time appointments, which means that the time they have available for research is reduced. Universities have been proliferating and many of them are far too small to be really viable, which further hinders university science. In addition, the universities are less productive than their counterparts in the industrial world.[36] In short, while the university is probably still a good location for much basic research, the research institute may also have much to offer.

In the United States research institutes have proliferated in the last twenty years or so, some of them in the university with connections with university departments, others as government or independent institutions. One reason for their emergence has been as a solution to the tensions felt between the scientist's teaching function and his research function in the university department environment. More important, the research institute allows a group of scientists to work together on a single or related set of problems, encouraging a concerted attack on a specific problem area of science.[37]

Given the need to provide constant interaction between different kinds of science (pure, applied) and different types of people (creators of science, users of science), the research institute may also be an ideal place for science in a developing country. It can offer greater interdisciplinary contacts, with fruitful results for the quality of research. The distinction between basic and applied science is more likely to break down in the research institute, with some scientists working together on problems of high predictability and feasibility, others in areas of less predictability, and others working alone on urgent problem areas of low predictability. As far as student training is concerned, the research institute need not necessarily be isolated from student education. Some research institutes will probably exist within a university structure, with members perhaps having formal membership in the appropriate university department. The research institute may participate in courses in the university department that lies within its research province, or offer entire, specialized courses to the student body, as did the Oswaldo Cruz Institute with its courses in microbiology. The research institute system also has the advantage of offering greater flexibility as to the timing of a student's introduction to research. A first year medical student might, for example, spend two or three months working in a research institute, aiding in a specific research program and acquiring middle-level research experience.

The other advantage of the research institute structure is that it may help bridge the gap between research and its applications. Very often in developing countries, potential users of research knowledge are unaware of where or what research is going on. The scientist in the university is divorced from the more practical world of the engineer and is not in a position to inform him. In the United States, the industrial entrepreneur took the lead in discovering the potential applications of science. In developing countries, in the absence of a large and varied national industrial class that is familiar with science, it may be the scientist who must perform this entrepreneurial role by working closely with industrialists, policy makers, or government officials, as the case may be, becoming aware of their more practical interests and informing them of what is going on in the research field. The scientist as entrepreneur was certainly one factor in the success of the Oswaldo Cruz Institute.

An interesting case study of the function of the research institute in promoting industrial applications comes from India. Y. Nayudamma describes the success of the Central Leather Research Institute in Madras, where it became one of the functions of the Institute to persuade potential users (mainly the small-scale cottage industry entrepreneurs) of the potential uses of research, while gathering information about the practical needs of the industralists.[38] Such an exchange, he argues, can only come about within a "pre-existing matrix of continuous and intimate interaction or dialogue between the research worker and the industrialists. The two must live in a common culture in which each can imagine himself in the role of the other."[39] The new US AID-São Paulo State contract in science and technology in Brazil attempts to create such a shared culture. State grants, matched by AID money, are to be awarded to entrepreneurs and industrialists for technical innovations, the grant awarded on condition that the entrepreneur works with the local research institute in his field, and the local research institute in its turn receives the aid of research institutions in the United States, designated by the São Paulo state government.[40]

In suggesting some of the advantages of the research institute for a developing country, I do not wish to ignore the universities, but to suggest a balance between the science of the university and that of the research institute. The main purpose of examining institutions in science with fresh eyes is to seek ways to establish a continuum between science and its uses that in the developing countries is less easily achieved than in the industrial countries, with their much larger research systems.

Conclusion

Looking to the future, one cannot be very optimistic about the possibility of closing the gap in material standards of living, conditions of health, and consumption of food that exists between the developing countries and the developed world. The closing of this gap depends on much more than the scientific and technical policy and the determination of the developing countries. Achievement of some kind of parity between countries would involve large-scale changes in the industrialized countries' consumption of food, energy, and water, changes that do not seem likely to come about in the near future. Nor am I particularly sanguine that the choices many of the developing countries will make about their own national science and technology efforts will be any better than those made in the industrial world. Science and technology for defense is likely to rank higher in the scale of priorities than health in many developing courtries, just as they do in Europe, Russia, and the United States. Industrial output, rather than a fair distribution of existing resources, will probably be the focus of national policy.

This pessimism is partly balanced, however, by a cautious optimism that in developing nations there is a new determination to forge autonomous paths of development, and to create at home the knowledge in science and technology that might help to achieve this goal.

References

[1]Joseph Ben-David, *The Scientist's Role in Society: A Comparative Study* (Englewood Cliffs, N. J.: Prentice-Hall, 1971), esp. pp. 123–129, 142–162.

[2]It was only in the 1930s that European physicists, fleeing fascism and invited to Brazil to work, came in sufficient numbers to have an impact on institutional physics. See James Rowe, "Science and Politics in Brazil: Background of the 1967 Debate on Nuclear Energy Policy," in Kalman Silvert (*ed.*), *The Social Reality of Scientific Myth* (New York: American Field Staff, Inc., 1969), pp. 91–122.

[3]Acuerdo de Cartagena, *Fundamentos para una polÍtica subregional de desarrollo tecnolÓgico* (Lima, Peru: Decimotercer periodo de sesiones ordinários de la Comisión, 12 de noviembre de 1973). Unpublished document of the Andean Pact, pp. 3–4.

[4]Developing countries spend about 0.2 percent of their GNP on research and development, although this estimate may be on the high side. According to the *World Plan of Action for the Application of Science and Technology to Development*, the target for expenditures should be 0.5 percent of GNP by the end of the decade, given the assumption of aggregate GNP growth rate of 6 percent per annum. See *Science in Developing Countries. World Plan of Action for the Application of Science and Technology to Development.* Abridgement of draft introductory statement by a group of consultants from the Institute of Development Studies and Science Policy Research Unit of University of Sussex. Advisory Committee on the Application of Science and Technology to Development. Reprinted in *Minerva* 9 (January 1971), 101-121.

[5]Some recent books that deal with science in the industrial world are: Joseph Ben-David, *The Scientist's Role in Society: A Comparative Study*, *op. cit.*, and his *Fundamental Research and the Universities; Some Comments on International Differences* (Paris: Organization for Economic Cooperation and Development, 1968); Jacques Ellul, *The Technological Society* (New York: Knopf, 1970); Christopher Freeman and Alison Young, *The Research and Development Effort in Western Europe, North America and the Soviet Union* (Paris: Organization for Economic Cooperation and Development, 1965); Daniel S. Greenberg, *The Politics of Pure Science* (New York: New American Library, 1967); Derek J. de Solla Price, *Little Science, Big Science* (New York: Columbia University Press, 1963); Don K. Price, *Government and Science; Their Dynamic Relation in American Democracy* (Oxford University Press, 1962) and *The Scientific Estate* (Cambridge, Mass.: The Belknap Press of Harvard University Press, 1965); Edward Shils, *The Criteria for Scientific Development: Public Policy and National Goals; A Selection of Articles from Minerva* (Cambridge, Mass.: MIT Press, 1968); Jerome R. Ravetz, *Scientific Knowledge and its Social Problems* (Oxford: Clarendon Press, 1971); Jean-Jacques Salomon, *Science and Politics* (Cambridge, Mass.: The MIT Press, 1973); and Alvin Weinberg, *Reflections on Big Science* (Cambridge, Mass.: The MIT Press, 1967).

[6]Jacob Schmookler, *Invention and Economic Growth* (Cambridge, Mass.: Harvard University Press, 1966), shows how technological innovation in four important sectors in the nineteenth century (railroads, agriculture, petroleum refining, and paper-making) occurred in response to economic demand and not as a result of scientific research.

[7]John J. Beer, *The Emergence of the German Dye Industry* (Urbana, Illinois: Illinois University Press, 1959). On the growth of relations between science and industry in the United States, see Kendall A. Birr, "Science in American Industry," in David D. Van Tassel and Michael G. Hall (*eds.*), *Science and Society in the United States* (Homewood, Illinois: The Dorsey Press, 1966), pp. 35–80.

[8]Several of the major industrial laboratories of the United States had been founded by 1900. By 1927 approximately 1,000 industrial research laboratories existed, employing about 19,000 people. By 1962 the number of scientists and engineers engaged in research and development in industry in the United States had reached 300,000. See Van Tassel and Hall, *op. cit.*, pp. 69–70.

[9]Jean-Jacques Salomon, *op. cit.*, pp. 51–52.

[10]Even on such a basic matter as the relation between science and economic growth, there is disagreement and uncertainty. Ben-David concludes, for example, that the relation between investment in science and economic growth is not unambiguously established "because thus far it has been impossible to measure whether the returns to the investor from applied research have been as great as some alternative uses of his capital." See Joseph Ben-David, "The Profession of Science and its Powers," *Minerva* 10 (July 1972), 379. Similarly, Harry Johnson, in "Some Economic Aspects of Science," *Minerva* 10 (January 1972), 10–18, writes that the solution to the problem of what science contributes to economic growth eludes him. Nonetheless, it is generally agreed that in the long run there are material and cognitive benefits to be derived from investments in science.

The phrase "scientific research system," describing the system of research and development found in the industrialized countries, is that of Salomon, *op. cit.*, Ch. 4.

[11]Derek J. de Solla Price argues that it is only in special, traumatic cases involving the breaking of the paradigms of science and technology that there is a direct flow from the research frontier of science to technology and vice versa. Much more usual is the flow from technology to science. See Derek J. de Solla Price, "Is Technology Historically Independent of Science? A Study in Statistical Historiography," *Technology and Culture* 6 (1965), 553–568. One example of direct flow is found in the field of quantum electronics; see Charles H. Townes, "Quantum Electronics and Surprise in the Development of Technology, The Problem of Research Planning," *Science* 159 (1968), 699–703. It may be that the chief effect of science on technology is that the training of scientists results in a large number of people who can enter the production process familiar with science. Edwin Layton, in "Mirror Image Twins: The Communities of Science and Technology", in George H. Daniels (*ed.*), *Nineteenth-Century American Science; A Reappraisal* (Evanston, Ill.: Northwestern University, 1972), pp. 210–230, describes the creation of a new technological community in the nineteenth century which allowed technological problems to be treated as scientific ones.

[12]J.-J. Salomon, *op. cit.*, p. 77.

[13]See Joseph Ben-David, *Fundamental Research and the Universities; Some Comments on International Differences*, *op. cit.*, pp. 56–61 for a discussion of this point. Ben-David emphasizes the absence of direct links between science and technology, and argues that it is above all the entrepreneurs who bring potential technological applications of science to the attention of potential users. He insists that there is no necessary continuum between science and applications.

[14]Harvey Brooks, Letter to *Minerva* 10 (April 1972), 327 and 328.

[15]A similar argument in favor of undertaking research within developing countries is made by Michael J. Moravcsik, "Technical Assistance and Fundamental Research in Underdeveloped Countries," *Minerva* 3 (Winter 1964), 197–209.

[16]Steven Dedijer, in one of the first articles devoted to the question of science in developing countries, gave the need to develop indigenous research capabilities top priority: "An objective estimate of the human and material resources available and necessary for the very first and each subsequent step in development demands the solution by scientists of a series of problems in statistics, demography, sociology, economics, geology, hydrology, geodesy, geography, etc. . . . Every aspect of national development policy depends on research conducted within the country, although it must, of course, be based on the achievements of, and conform with, the standards of international science." See "Underdeveloped Science in Underdeveloped Countries," *Minerva* 2 (August 1963), 64.

[17]For this figure see Christopher Freeman and A. Young, *op. cit.*, p. 66.

[18]See A. Hunter Dupree, *Science in the Federal Government: A History of Policies and Activities to 1940* (Cambridge, Mass.: The Belknap Press of Harvard University Press, 1957) pp. 258–263, for the failure of American research into yellow fever. See also Nancy Stepan, "Social Factors in Scientific Discovery: The Case of Yellow Fever," unpublished manuscript, for an analysis of the reasons behind the lack of American interest in research into the causes of yellow fever between 1870 and 1900.

[19]This point is made by a number of people. See for example, Alvin M. Weinberg, "Scientific Choice and Biomedical Science," *Minerva* 4 (Autumn 1965), 3–14. Salomon

also writes, "There is no longer any break in the process from the extension of pure knowledge to the creation of new families of technical objects. It is as hard to distinguish between pure research and applied research" and "The only distinction which still seems defensible is that based on the different channels of information and communication appropriate to science and technology." See J.-J. Salomon, *op. cit.*, pp. 83 and 80, respectively.

[20]J.-J. Salomon, *op. cit.*, p. 90.

[21]For a discussion of the need for local teams of science in the production of high-yield varieties of plants, see V. W. Ruttan and Yujrio Hayami, "Technology Transfer and Agricultural Development," *Technology and Culture* 14 (1974), 119-151.

[22]This point is made by Alvin M. Weinberg, "Scientific Choice and Biomedical Science," *op. cit.*

[23]It is not mentioned, for example, by George Basalla, in his article, "The Spread of Western Science," *Science* 156 (May 1967), 611-622.

[24]Derek J. de Solla Price, "Is Technology Historically Independent of Science? A Study in Statistical Historiography," *op. cit.*

[25]The literature on dependency is already large and is growing fast. Here I cite some books and articles that are useful for the analysis of science: Constantino V. Vaitsos, *Comercialización de tecnología en el Pacto Andino* (Lima, Peru: Instituto de Estudios Peruanos, 1973), and his "Power, Knowledge and Development Policy: Relations Between Transnational Enterprises and Developing Countries," The 1974 Dag Hammerskjold Seminar on the Third World and International Economic Change (Uppsala, Sweden: The Dag Hammarskjöld Foundation, 1974), unpublished manuscript; O. Sunkel, "Underdevelopment, the Transfer of Science and Technology, and the Latin American University," *Human Relations* 24 (1971), 1-18; Francisco R. Sagasti, "Discussion Paper: Underdevelopment, Science and Technology: The Point of View of the Underdeveloped Countries," *Science Studies* 3 (1973), 47-59; José Leite Lopes, *La ciencia y el dilema de América Latina: dependencia o liberación* (Buenos Aires: Siglo Ventiuno Argentina Editores, 1972); Amilcar O. Herrera, *Ciencia y política en América Latina* (Buenos Aires: Siglo Ventiuno Editores, 1971); Miguel S. Wionczek, *Inversión y tecnología extranjera en América Latina* (Mexico: Editorial de Joaquín Mortiz, 1971).

Dependency in science has also been discussed as a "center-periphery" problem. At different times, the center of a given discipline in science is located in one country rather than another. Periphery countries send their students to the center for training, and measure themselves by the standards of the center. They take steps to improve their position vis-à-vis the center. According to this analysis, dependency is resolved by the replacement of the center by the periphery. Thus there is a continual shift in the location of the center. See Rainald von Gizycli, "Center and Periphery in the International Scientific Community: Germany, France and Great Britain in the 19th Century," *Minerva* 11 (October 1973), 474-494. In most dependency theory writings, however, relations of dependency are fixed and cannot be altered without revolutionary changes in capitalistic economy.

[26]Albert O. Hirschman, *A Bias for Hope: Essays on Development and Latin America* (New Haven and London: Yale University Press, 1971), p. 93.

[27]*Ibid.*, pp. 93–94.

[28]This may mean that even when the host country possesses the know-how for two-thirds of a particular technological package it still has to purchase this know-how because it is part of an entire package. The Andean Pact, and such important national industries as the Peruvian national oil company (Petroperu) have identified the problem of "disaggregating" technology packages as a major technological goal for the near future.

See Constantino V. Vaitsos, *Comercializacibn de tecnologla en el Pacto Andino, op. cit.*, passim, for a discussion of some of the hidden costs of technology transfer. Adaptation of patents is another problem. Very often the adaptation reverts back to the original patentor, so that while the costs of technology adaptation accrue to the host country, the benefits do not. See Jorge M. Katz, "Industrial Growth, Royalty Payments and Local Expenditure on Research and Development," in Victor L. Urquidi and Rosemary Thorp, *Latin America in the International Economy; Proceedings of a Conference Held by the International Economic Association in Mexico City, Mexico* (New York: MacMillan, 1973), pp. 197–223, and his *Importacibn de tecnologla, aprendizaje local e industrializacibn dependiente* (Buenos Aires; Instituto Torcuata di Tella, Centro de Investigacionas Econ6micas, 1972).

[29]This cycle of effects are described in Acuerdo de Cartagena, *op. cit.*

[30]The conclusion of the "dependency" view of technology is that the "knowledge needed to buy knowledge" must be built up in the developing countries as a first step. For this latter expression, see Carlos F. Diáz-Alejandro, "North-South Relations: The Economic Component," Economic Growth Center Discussion Paper, No. 200, Yale University, April 1974. Already several countries in Latin America have taken measures to extract more "knowledge to create knowledge" in their dealing with foreign companies. Particularly interesting is the attempt by the Andean Pact countries to build up industrialization while lessening technological and scientific dependence. Through a regional association and agreement on bargaining strategies, they hope that their power to select, adapt and invent new technical solutions to the problems of the subregion will be increased. The mechanisms proposed to achieve this are varied. In return for access to markets, they hope to involve foreign firms more deeply in local research and development, to stimulate rather than throttle the employment of national scientists and technicians, to learn management skills from multinational corporations—in short, to use multinationals to create linkages between national efforts in science and technology and the industrial process. These mechanisms are described in various decisions of the Andean Pact, such as By-Laws for the Common Treatment of Foreign Capital, Trademarks, Patents, Licensing Agreements and Royalties, as stated in Decisions 24, 37, and 37a of the Cartagena Agreement.

The new international demand for primary products, such as oil, may help the developing countries in their attempt to extract more from foreign technology by increasing their bargaining power vis-à-vis foreign companies. Few doubt that Latin American countries will continue to rely on foreign technology. Indeed, this reliance will certainly increase as industrialization proceeds. More work needs to be done on the possible benefits of technological purchase under new, less dependent conditions.

[31]Claire Nader, "Technical Experts in Developing Countries," in Claire Nader and A. Zahlan, (*eds.*), *Science and Technology in Developing Countries.* Proceedings of a

Conference Held at the American University of Beirut, Lebanon, 27 November–2 December, 1967 (Cambridge: Cambridge University Press, 1969), p. 451.

[32]Michael Polanyi, "The Republic of Science," *Minerva* 1 (Autumn 1962), 54–73. The opposite view is taken by C. F. Carter in "The Distribution of Scientific Effort," *Minerva* 1 (Winter 1963), 172–181.

[33]Alvin M. Weinberg, "Criteria for Scientific Choice," *Minerva* 1 (Winter 1963), 159–171.

[34]Gerald Gordon, Ann E. MacEachron, G. Lawrence Fisher, "A Contingency Model for the Design of Problem-Solving Research Programs: A Perspective on Diffusion Research," *Health and Society* 52 (Spring 1974), 185–220.

[35]On the significance of the university for science, see Joseph Ben-David, "Review Article. Scientific Growth: A Sociological View," *Minerva* 2 (Summer 1964), 455–476, and his *Fundamental Research and the Universities; Some Comments on International Differences, op. cit.*

[36]Many aspects of the Latin American university are discussed in *Education, Human Resources and Development in Latin America* (United Nations, New York: Economic Commission for Latin America, 1968).

[37]See Peter H. Rossi, "Researchers, Scholars and Policy Makers: The Politics of Large Scale Research," *Daedalus* 93 (Fall 1964), 1142–1161.

[38]Y. Nayudamma, "Promoting the Industrial Application of Research in an Underdeveloped Country," *Minerva* 5 (Spring 1967), 323–339.

[39]*Ibid.*, p. 324.

[40]Interview with Dr. Robert Goeckerman, Department of State, U.S.A.

Selected Bibliography

1 *Manuscript Sources*

The main manuscript sources used in this book are those relating to the history of the Oswaldo Cruz Institute, found in the Institute in Rio de Janeiro. I have classified these sources in the following way: (1) *The Oswaldo Cruz Files*—Twenty-five files of handwritten documents, many in Cruz' hand, describing different aspects of the institute's work, together with an excellent collection of photographs, and much of the official correspondence (oficios) between the institute and various government agencies; (2) *Administrative Records*—Handwritten records, contained in a series of volumes, describing the appointment, terms of employment, and activities of the staff members of the institute housed in the Administrative Section of the institute; and (3) *Museum Documents*—A series of miscellaneous documents, discovered at the institute, also describing various activities of the institute. These sources were consulted with the permission of the then Director of the Oswaldo Cruz Institute, Dr. Francisco de Paulo da Rocha Lagoa.

Although this work did not pretend to study the biographical sources of Oswaldo Cruz in depth, two sources that were useful for Cruz' work at the institute were: (1) *The Oswaldo Cruz Filho Family Files*—Files of newspaper clippings and photographs describing Cruz' career as Director of the Public Health Department between 1903 and 1909. My thanks to Dr. Oswaldo Cruz Filho for permission to consult this source; (2) Letters from Oswaldo Cruz to Henrique da Rocha Lima—These letters were sent to Rocha Lima while the latter was abroad, and they reflect some of Cruz' concerns about the institute. The first letter in the collection dates from 1901, and the last in the series is dated 1915. These letters were examined by permission of Mrs. Henrique da Rocha Lima.

Other manuscript sources consulted in this study were: (1) Some correspondence received by Dr. Emílio Ribas while Director of the Sanitary Services of São Paulo state after 1898—This correspondence is collected in typewritten copies in São Paulo, Secretário de Saúde e Assistência Social. Correspondência de Emílio Ribas (setembro 1964); (2) Handwritten reports on the work of the Bacteriological Institute of São Paulo, in Adolfo Lutz' hand—Many of these were never published by the state. See São Paulo, Instituto Adolfo Lutz, Departmento de Administração, *Relatórios* for years 1895, 1896, 1899, 1900, 1901, 1902, 1903, and 1906; (3) The handwritten original of Professor Martin Ficker's report on the Bacteriological Institute of São Paulo, written in 1913 and not published at the time—See São Paulo, Instituto Adolfo Lutz, Departmento de Administração. Martin Ficker,

Programma para a reorganização do Instituto Bacteriológico, 1913; (4) São Paulo, Instituto Adolfo Lutz, Departmento de Administração—Vital Brasil, Realtório sôbre a peste bubônica em Santos, apresentado ao Dr. Director do Instituto Bacteriológico em 27 de novembro de 1899. Handwritten original; (5) Unpublished Reports (Memórias históricas) of the Medical School of Rio de Janeiro—Professor Francisco Bruno Lobo is preparing them for publication and kindly lent me a number of items in mimeographed form. They are cited under Lobo, Francisco Bruno, in the bibliography.

2 *Official Publications*

A. Federal Government

Brazil. *Anuário de estatística do Brasil,* 1966.

———. *Colecção das leis da República dos Estados Unidos do Brasil de 1908.*

———. Congresso Nacional. *Annaes da Câmara dos Deputados. Sessões* de 1903, Vols. I, III, V, VI, VII, and VIII. *Sessões* de 1906, Vols. III, and VII.

———. ———. Directoria Geral de Saúde Pública. *Annuário de estatística demographosanitária,* 1903, 1904, 1905, 1906, 1907, 1908, 1909, 1910, 1911.

———. ———. Placido Barbosa e Cassio Barboso de Rezende, *Os serviços de saúde pública no Brasil, especialmente na cidade do Rio de Janeiro de 1808 à 1907 (Esboço histórico e legislação)* 2 vols. (Rio de Janeiro: Imprensa Nacional, 1909).

———. *Relatório apresentado ao Exm. Sr. Dr. J.J. Seabra, Ministro da Justiça e Negocios Interiores, pelo Dr. Oswaldo Gonçalves Cruz, Director Geral de Saúde Pública,* 1904, 1905, 1906, 1907, 1908.

———. Ministério da Fazenda. *Leis do orçamento de receita e despesa para o exercício de 1901, 1902, 1903, 1904, 1905, 1906, 1907, 1908, 1909, 1910, 1911, 1912, and 1913.*

B. State Government

São Paulo. *Actos do poder legislativo de Estado de São Paulo* (São Paulo: Diário Official, 1892).

———. *Leis e resoluções decretos pelo Congresso Legislativo do Estado de São Paulo em 1891* (São Paulo: Diário Official, 1892).

———. *Collecção das leis e decretos do Estado de São Paulo de 1893,* 1894, 1895, 1896, 1897, 1898, 1899, 1900, 1901, 1902, 1903, 1904.

———. Congresso. Câmara dos Deputados do Estado de São Paulo. *Annaes das sessões de 1891,* 1892, 1893.

———. Directoria do Serviço Sanitário. *Annuário estatístico da secção demographia. Anno de 1897,* 1898, 1899, 1900, 1904, 1907.

———. Instituto Bacteriológico, *Coletanêa de trabalhos do Instituto Bacteriológico, 1895–1933.*

———. Instituto de Butantã. *Coletanêa de trabalhos, 1901–1917* (São Paulo: Diário Official, 1918).
———. Secretário de Estado dos Negocios do Interior. *Relatório apresentado ao Presidente do Estado de São Paulo pelo Secretário dos Negocios do Interior, 1893,* 1894, 1895, 1896.
———. Serviço Sanitário do Estado de São Paulo. Instituto Bacteriológico. Bruno Pestana, "A febre typhoide em São Paulo."

3. *Books and Articles*

Academia Nacional de Medicina. *Em commemoração do centenário do ensino médico* (Rio de Janeiro: Jornal do Commercio, 1908).
Agassiz, Louis, and Agassiz, Elizabeth C. *A Journey in Brazil* (Boston: Ticknor and Fields, 1868).
Affonso, Barão Pedro. *Relatórios dos trabalhos do Instituto Vaccínico do Distrito Federal, segundo de um retrospetivo dos trabalhos vaccínicos de 1887 à 1917* (Rio de Janeiro: n.p., 1917).
Albuquerque, M. Sabina de. "Dr. Aldolfo Lutz," *Revista do Instituto Adolfo Lutz* 10 (1950), 9–30.
Alves, Francisco de Paulo Rodrigues, Filho, and Rodrigues, Oscar, eds. *Centenário do Conselheiro Rodrigues Alves,* 2 vols. (São Paulo: Empresa Gráfica da "Revista dos Tribunais" Ltda., 1951).
Amaral, Afranio do. *Animaes venenosos do Brasil* (São Paulo: Instituto Butantan, 1930).
Andrade, Nuno de. *Febre amarella e o mosquito* (Rio de Janeiro: Jornal do Commercio, 1903).
Aragão, Henrique de Beaurepaire. "Sôbre o cyclo evolutivo do halterídio do pombo (nota preliminar)," *Brasil-Médico* 21 (1907), 141–142, 301–303.
———. *Sôbre o cyclo evolutivo do halterídio do pombo (1ª e 2ª notas)* (Rio de Janeiro: Besnard Frères, 1907).
———. *Oswaldo Cruz e a escola de Manguinhos.* Conferência realizada no Centro Acadêmico Oswaldo Cruz de São Paulo em 20 de septembro de 1940. Segunda edição (Rio de Janeiro: Imprensa Nacional, 1945).
———. *Notícia histórica sôbre a fundação do Instituto Oswaldo Cruz (Instituto de Manguinhos)* (Rio de Janeiro: Serviço Gráfico do Instituto Brasileiro de Geografia e Estatística, 1950).
———. "Carlos Chagas, diretor do Manguinhos," *Memorias do Instituto Oswaldo Cruz* 51 (1953), 1–10.
Artz, Frederick B. *The Development of Technical Education in France, 1500–1850* (Cambridge, Massachusetts: Cleveland Society for the History of Technology, 1966).
Austregesílio, Antônio. *Oswaldo Cruz: vida gloriosa de Oswaldo Cruz* (Rio de Janeiro: Departmento Nacional de Saúde, 1937).

Azevedo, Fernando de. *Brazilian Culture: An Introduction to the Study of Culture in Brazil.* Translated by William Rex Crawford (New York: The Macmillan Company, 1950).

———. *As ciências no Brasil*, 2 vols. (Rio de Janeiro: Edições Melhoramentos, 1955).

Bacellar, Renato Clark. *Brazil's Contribution to Tropical Medicine and Malaria: Personalities and Institutions.* Translated by Anita Farquhar (Rio de Janeiro: Gráfica Olympica Editôra, 1963).

Basalla, George. "The Spread of Western Science," *Science* 156 (5 May, 1967), 611-622.

———. ed. *The Rise of Modern Science: Internal or External Factors?* (Lexington, Massachusetts: D.C. Heath and Company, 1968).

Bates, Henry Walter. *The Naturalist on the River Amazons* (Berkeley and Los Angelos: University of California Press, 1962).

Beer, Joseph. *The Emergence of the German Dye Industry* (Urbana, Illinois: Illinois University Press, 1959).

Bello, José Maria. *A History of Modern Brazil, 1889-1964* (Stanford, California: California University Press, 1966).

Ben-David, Joseph. "Review Article. Scientific Growth: A Sociological View," *Minerva* 2 (Summer 1964), 455-476.

———. *Fundamental Research and the Universities; Some Comments on International Differences* (Paris: OECD, 1968).

———. *The Scientist's Role in Society. A Comparative Study* (Englewood Cliffs, New Jersey: Prentice-Hall, Inc., Foundation of Modern Sociology Series, 1971).

———. "The Profession of Science and its Powers," *Minerva* 10 (July 1972), 362-383.

Bernard, P.-Noel, ed. *Les Instituts Pasteurs d'Indochine* (Centenaire de Louis Pasteur, 1822-1895) (Saigon: Imprimerie Nouvelle Albert Portail, 1922).

Bonner, Thomas N. *American Doctors and German Universities, A Chapter in International Relations, 1870-1914* (Lincoln: University of Nebraska Press, 1963).

Borgmeier, T. "Artur Neiva: a propósito do seu sexagésimo anniversário natálico," *Revista de Entomologia* 2 (June 1940) (Volume commemorativo do 60° anniversário natálico de Artur Neiva).

Boxer, Charles R. *The Golden Age of Brazil, 1695-1750: Growing Pains of a Colonial Society* (Berkeley and Los Angeles: University of California Press, 1964).

———. *The Portuguese Seaborne Empire, 1415-1825* (New York: Alfred A. Knopf, 1969).

Branner, John Casper. *A Bibliography of the Geology, Minerology, and Paleontology of Brazil* (Rio de Janeiro: Imprensa Nacional, 1903).

———. "Memorial of Orville A. Derby," *Bulletin of the Geological Society of America* 27 (1916), 15-21.

Brasil, Vital. "A peste bubônica em Santos. Trabalho do Instituto Bacteriológico de São Paulo, apresentado ao director do Instituto Bacteriológica, 27 de novembro, 1899," *Revista Médica de São Paulo* 2 (1899), 343-355.

———. *La défense contre l'ophidisme.* Traduction française par le Professeur J. Maibon. 2 ed. (St. Paul: Pocai-Weiss e C., 1914).

———. *Memória histórica do Instituto de Butantan* (São Paulo: Elvino Pocai, 1941).

Brazilian Research Council, National Academy of Sciences. *Science and Brazilian Development, Report of a Workshop on the Contribution of Science and Technology to Development* (Brazilian Research Council, National Academy of Sciences, National Research Council, in Cooperation with the Agency for International Development, 1966).

Briquet, R.C. "Adolfo Lutz: exemplo e gloria da ciência médica brasileira," *Revista do Instituto Adolfo Lutz* 1 (1941), 203-216.

Burnham, John C. *Science in America: Historical Selections* (New York: Holt, Rinehart and Winston, Inc., 1971).

Burns, E. Bradford. "The Role of Azeredo Coutinho in the Enlightenment of Brazil," *Hispanic American Historical Review* 44 (May 1964), 145-160.

Burton, Isabel. *The Life of Captain Sir Richard F. Burton*, 2 vols. (London: Chapman and Hall, Ltd., 1893).

Burtt, Edwin A. *The Metaphysical Foundations of Modern Physical Science: A Historical and Critical Essay* (London: Routledge and Kegan Paul, 1949).

Bustamente, Emília. *As bibliotecas especializadas como fontes de orientação na pesquisa científica* (Rio de Janeiro: Instituto Oswaldo Cruz, 1958).

Cajori, Florian. *The Early Mathematical Sciences in North America and South America* (Boston: R. G. Badger, 1928).

Calmette, A. "Institut Pasteur," *in* France, Ministère du Travail, *Livre d'or de la commémoration nationale du centenaire de la naissance de Pasteur* (Paris: Imprimerie Nationale, 1928).

———. "Les missions scientifiques de l' Institut Pasteur et l'expression coloniale de la France," *Revue Scientifique* 89 (1912), 129.

———. "Pasteur et les Instituts Pasteur," *Revue d'Hygiene* 45 (1923), 385.

Calmón, Pedro. *História social do Brasil*, 3 vols. (São Paulo: Companhia Editôra Nacional, 1937-40).

———. *História do Brasil.* 2 ed, 7 vols. (Rio de Janeiro: José Olympio, 1963).

Calógeras, João Pandiá. *A History of Brazil* (New York: Russell and Russell, Inc., 1963).

Cardwell, D.L.A. *The Organization of Science in England: A Retrospect* (Melbourne: Heinemann, 1957).

Chagas, Carlos. "Prophylaxia do impaludismo," *Brasil-Médico* 20 (1906), 315-317, 337-340, 419-422 and *Brasil-Médico* 21 (1907), 151-154.

———. "Prophylaxia do impaludismo," *Revista Médica de São Paulo* 11 (1908), 391-399.

———. "Nova tripanozomiaze humano. Estudos sôbre a morfolojia e o ciclo evolutivo do Schistzotrypanum cruzi, n. gen., n. sp., ajente etiolójico de nova entidade mórbida do homem," *Memorias do Instituto Oswaldo Cruz* 1 (1909), 159–218.

———. "Neue Trypanosomen. Vorlaufige mitteilung," *Archiv für Schiffs-und-Tropenhygiene* 13 (1909), 120–122.

———. "Descoberta do Tripanosoma cruzi e verificação da Triponosomiase americana. Retrospecto histórico," *Memorias do Instituto Oswaldo Cruz* 15 (1922), 67–76.

———. "Adolpho Lutz," *Memorias do Instituto Oswaldo Cruz* 18 (1925), i–xxii.

Chagas Filho, Carlos. "Carlos Chagas, 1879–1934," *O Hospital* 54 (Julho 1958), 9–15.

Clagett, Marshall. *The Science of Mechanics in the Middle Ages* (Madison: University of Wisconsin Press, 1959).

Cline, Howard F. "The *Relaciones Geográficas* of the Spanish Indies, 1577–1586," *Hispanic American Historical Review* 44 (August 1964), 341–374.

Cobo, Bernabé. *História del nuevo mundo.* Pub. por primera vez con notas y otras ilustraciones de Marcos Jiménez de la Espada, 4 vols. (Sevilla: Imp. de E. Rasco, 1890–93).

Coni, Antônio Caldas. *A escola tropicalista bahiana: Paterson, Wucherer, Silva Lima* (Bahia: Tipografia Beneditina Ltda., 1952).

Cooper, Donald. "Oswaldo Cruz and the Impact of Yellow Fever on Brazilian History," *The Bulletin of the Tulane University Medical Faculty* 26 (February 1967), 49–52.

Couty, Louis. Musée National, *Cours de biologie expérimentale.* Leçon d'ouverture (Rio de Janeiro: G. Leuzinger e Fils, 1880).

———. "O ensino superior no Brasil," *Gazeta Médica da Bahia* 15 (1884), 521–532.

———. "Os estudos experimentaes no Brasil," *Revista Brasileira*, Primeiro anno (1884), 215–239.

Crombie, A.C. *Robert Grosseteste and the Origins of Experimental Science, 1100–1700* (Oxford: Clarendon Press, 1953).

———. ed. *Scientific Change; Historical Studies in the Intellectual Social and Technical Conditions for Scientific Discovery and Technical Invention, from Antiquity to the Present* (New York: Basic Books, 1963).

Crosland, Maurice P. *The Society of Arcueil: A View of French Science at the Time of Napoleon I* (Cambridge, Mass.: Harvard University Press, 1967).

Corrêa Filho, Virgilio. *Alexandre Rodrigues Ferreira: vida e obra do grande naturalista brasileiro* (São Paulo: Companhia Editôra Nacional, Bibliotheca pedagógica brasileira, Ser. 5a, Brasiliana 144, 1933).

Cruz, Oswaldo. *Vehiculação microbiana pelas aguas.* These apresentada à Faculdade de Medicina do Rio de Janeiro, em 8 de novembro de 1892 (Rio

de Janeiro: Papelaria e Impressora, 1893).
———. *Relatório acêrca da moléstia reinante em Santos, apresentado pelo Dr. Oswaldo Gonçalves Cruz à S. Ex. o Sr. Ministro da Justiça e Negocios e Interiores* (Rio de Janeiro: Imprensa Nacional, 1900).
———. *A vaccinação anti-pestosa.* Trabalho do Instituto Sôrotherápico Federal do Rio de Janeiro (Instituto de Manguinhos) (Rio de Janeiro: Besnard Frères, 1901).
———. *Dos accidentes em sôrotherapia.* Trabalho do Instituto Sôrotherápico Federal do Rio de Janeiro (Instituto de Manguinhos) (Rio de Janeiro: Besnard Frères, 1902).
———. "Resumo da memória apresentada pelo Delegado do Brazil à 3ª convenção sanitária internacional reunida na cidade México de 2 à 7 de dezembro de 1907," in Oswaldo Cruz, *Opera omnia*, pp. 527–533.
———. *Prophylaxia da febre amarella. Memória apresentada ao 4° Congresso Médico Latino-Americano* (Rio de Janeiro: Jornal do Commercio, 1909).
———. "The Sanitation of Rio," *The Times*, Dec. 28, 1909, *in* Oswaldo Cruz, *Opera omnia*, pp. 556–562.
———. *Considerações gerais sôbre as condições sanitárias do Rio Madeira* (Rio de Janeiro: Papeleria Americana, 1910).
———. "Prophylaxis of Malaria in Central and Southern Brazil," *in* Ronald Ross, *The Prevention of Malaria* (London: John Murray, 1910).
———. *Relatório sôbre as condições medico-sanitárias do Valle do Amazonas* (Rio de Janeiro: Jornal do Commercio, 1913).
———. *Discurso pronunciado na Academia Brasileira de Letras* (26 de junho de 1913) (Rio de Janeiro: Röhe, 1913).
———. *Opera omnia* (Rio de Janeiro: Impressora Brasileira, Ltda., 1972).
Cruz Costa, João. *Contribuição a história das idéias no Brasil (O desenvolvimento da filosofia no Brasil e a evolução histórica nacional)* (Rio de Janeiro: José Olympio, 1956).
———. *O positivismo na República: notas sôbre a história do positivismo no Brasil* (São Paulo: Companhia Editôra Nacional, Biblioteca pedagógica brasileira, Ser. 5a, Brasiliana 291, 1956).
Cutright, Paul R. *The Great Naturalists Explore South America* (New York: The Macmillan Company, 1940).
Daniels, George H. "The Process of Professionalization in American Science: The Emergent Period, 1820–1860," *Isis* 58 (Summer 1967), 151–166.
———. *American Science in the Age of Jackson* (New York: Columbia University Press, 1968).
———, ed. *Nineteenth Century American Science; A Re-appraisal* (Evanston, Illinois: Northwestern University Press, 1972).
Dedijer, Steven. "Underdeveloped science in underdeveloped countries," *Minerva* 2 (August 1963), 61–81.
Delaunay, Albert. *L'Institut Pasteur, des origines à aujourd'hui* (Paris: Editions

France-Empire, 1962).

Derby, Orville A. "The Present State of Science in Brazil," *Science* 1 (1883), 211–214.

Dias, Ezequiel Caetano. *O Instituto Oswaldo Cruz: resumo histórico 1899–1918* (Rio de Janeiro: Manguinhos, 1918).

———. *Traços biográficos de Oswaldo Cruz* (Rio de Janeiro: Imprensa Nacional, 1945).

Díaz-Alejandro, Carlos F. "North-South Relations: The Economic Component," Economic Growth Center Discussion Paper, No. 200, Yale University, April 1974.

Dos Santos, Ruy. "Ruy Barbosa e o médico," *Anais Paulistas de Medicina e Cirurgia* 59 (1950), 295–305.

Dundas, Robert. *Sketches of Brazil, including New Views on Tropical and European Fever, with Remarks on a Premature Decay of the System Incident to Europeans on their Return from Hot Climates* (London: John Churchill, 1852).

Dupree, A. Hunter. *Science and the Federal Government: A History of Policies and Activities to 1940* (Cambridge, Mass.: The Belknap Press of Harvard University Press, 1957).

———. *Asa Gray, 1810–1888* (New York: Atheneum, 1968).

Ellul, Jacques. *The Technological Society* (New York: Knopf, 1970).

Falcão, Edgard de Cerqueira. *Martin Heinrich Karl Lichtenstein, estudo crítico dos trabalhos de Marcgrave e Piso sôbre a história natural do Brasil à luz dos desenhos originais* (São Paulo: Brasiliensia Documenta 2, 1961).

———. *Gaspar de Oliveira Vianna 1895–1914. Opera omnia* (São Paulo: A Gráfica de 'Revista das Tribunais,' 1962).

———. *Oswaldo Cruz; monumenta histórica*, 3 vols. (São Paulo: Brasiliensia Documenta 6, 1971–73).

Ferreira, Alexandre Rodrigues. *Viagem filosófica às capitanias do Grão-Pará, Rio Negro, Mato Grosso e Cuiabá. Memórias: zoologia e botânica* (Rio de Janeiro: Conselho Federal de Cultura, 1972).

Fleming, Donald H. *William H. Welch and the Rise of Modern Medicine* (Boston: Little, Brown, 1954).

Flexner, Abraham. *Medical Education in the United States and Canada; Report to the Carnegie Foundation for the Advancement of Teaching* (New York: Carnegie Foundation for the Advancement of Teaching, Bulletin No. 4, 1910).

Fonseca Filho, Olympio da. *Primórdios da microscópia óptica no Brasil: alguns pioneiros.* Unpublished manuscript. No date.

———. "O Brasil e as ciências naturais nos séculos XVI à XVIII," *Ciência e Cultura* 25 (1973), 946–1029.

Fort, Joseph Auguste. *Le récit de ma vie avec la description d'un voyage et d'un séjour dans l'Amérique du Sud. Autobiographie* (Paris: L. Bataille, 1893).

Fosdick, Raymond Blaine. *The Story of the Rockefeller Foundation* (New York: Harper, 1952).

Fraga, Clementino. *A febre amarella no Brasil: notas e documentos de uma grande campanha sanitária* (Rio de Janeiro: Officina Gráphica da Inspectoria de Demographia Sanitária, 1930).

———. *Vida e obra de Osvaldo Cruz* (Rio de Janeiro: José Olympio, 1972).

Freeman, Christopher, and Young, Alison. *The Research and Development Effort in Western Europe, North America and the Soviet Union; An Experimental Comparison of Research Expenditures and Manpower in 1962* (Paris: OECD, 1965).

Gale, A.H. *Epidemic Diseases* (London: Penguin Books, 1959).

Garrett, Lorraine Williams. *The Snake Farm at Butantan, Brazil* (Washington, D.C.: Pan American Union, with the Co-operation of the Office of the Coordinator of Inter-American Affairs, 1942).

Gizycki, Rainald von. "Center and Periphery in the International Scientific Community: Germany, France and Great Britain in the 19th Century," *Minerva* 11 (October 1973), 474–494.

Godoy, Alcides. "Nova vacina contra o carbúnculo sintomático," *Memorias do Instituto Oswaldo Cruz* 2 (1910), 11–21.

Goldsmith, Maurice, and Mackay, Alan, eds. *The Science of Science* (London: Pelican Books, 1964).

Gonsalves, Alpheu Diniz, ed. *Orville Derby's Studies on the Paleontology of Brazil: Selection and Coordination of this Geologist's Out of Print and Rare Works* (Rio de Janeiro: Published under the Direction of the Executive Commission for the First Centenary Commemorating the Birth of Orville A. Derby, 1952).

Goodman, Edward J. *The Explorers of South America* (New York: The Macmillan Company, 1972).

Graham, Richard. *Britain and the Onset of Modernization in Brazil 1850–1914* (Cambridge: Cambridge University Press, 1968).

Greenberg, Daniel S. *The Politics of Pure Science* (New York: New American Library, 1967).

Guenther, Konrad. *A Naturalist in Brazil: The Record of a Year's Observation of her Flora, her Fauna, and her People.* Translated by Bernard Miall (Boston and New York: Houghton Mifflin Company, 1931).

Guerra, E. Sales. *Osvaldo Cruz* (Rio de Janeiro: Casa Editôra Vecchi Limitada, 1940).

Guerra, Francisco. *Bibliografia médica brasileira, período colonial 1808–1821* (New Haven, Connecticut: Yale University School of Medicine, 1958).

Haring, C.H. *Empire in Brazil. A New World Experiment with Monarchy* (Cambridge, Massachusetts: Harvard University Press, 1958).

Hartt, C. Frederick. *Geology and Physical Geography of Brazil* (Boston: Fields, Osgood and Co., 1870).

Herrera, Amilcar O. *Ciencia y política en América Latina* (Buenos Aires: Siglo Ventiuno Editores, 1971).

Hilton, Ronald. *The Scientific Institutions of Latin America* (Stanford: California Institute of International Studies, 1970).

Hindle, Brooke. *The Pursuit of Science in Revolutionary America, 1735–1789* (Chapel Hill: University of North Carolina Press, 1956).

Hirschman, Albert O. *The Strategy of Economic Development* (New Haven: Yale University Press, 1958).

———. *A Bias for Hope: Essays on Development and Latin America* (New Haven and London: Yale University Press, 1971).

Howarth, Osbert J.R. *The British Association for the Advancement of Science: A Retrospect 1831–1921* (London: The Association, 1922).

Ihering, Hermann von. *The Anthropology of the State of São Paulo* (São Paulo: Duprat and Co., 1904).

Instituto Histórico e Geográphico Brasileiro. *Diccionário histórico geográphico e ethnográphico do Brasil*, 2 vols. (Rio de Janeiro: IHGB, 1922).

Instituto Oswaldo Cruz, *Lista cronolójica das publicações do Instituto Oswaldo Cruz de 1900 à 1915* (Rio de Janeiro: Manguinhos, n.d.).

———. *Cinqüentenário da descoberta da doença de Chagas. Carlos Chagas (1879–1934), Bio-bibliografia* (Rio de Janeiro: Instituto Oswaldo Cruz, Biblioteca, 1959).

Johnson, Harry. "Some Economic Aspects of Science," *Minerva* 10 (January 1972), 10–18.

Katz, Jorge M. *Importación de tecnología, aprendizaje local e industrialización dependiente* (Buenos Aires: Instituto Torcuata di Tella, Centro de Investigaciones Económicas, 1972).

———. "Industrial Growth, Royalty Payments and Local Expenditure on Research and Development," in Urquidi, Victor L., and Thorp, Rosemary, eds. *Latin America in the International Economy* (New York: Macmillan, 1973).

Kett, Joseph F. *The Formation of the American Medical Profession; The Role of Institutions, 1780–1860* (New Haven: Yale University Press, 1968).

Koyré, Alexandre. *From the Closed World to the Infinite Universe* (Baltimore: Johns Hopkins Press, 1957).

Lacerda, João Batista de. *Fastos do Museu Nacional do Rio de Janeiro: recordações históricas scientíficas fundadas em documentos authênticos e informações verídicas.* Obra executada por indacação e sob o patronato do Sr. Ministro de Interior, Dr. J. J. Seabra (Rio de Janeiro: Imprensa Nacional, 1905).

Lacorte, J. Guilherme. "A atuação de Oswaldo Cruz no aparecimento da peste bubônico no Brasil," *A Fôlha Médica* 54 (Fevereiro 1967), 183–188.

Lambert, Royston. *Sir John Simon, 1816–1904, and English Social Administration* (London: MacGibbon and Kee, 1963).

Landa, Diego de. *Landa's Relación de las cosas de Yucatán*, a translation, edited with notes by Alfred M. Tozzer (Cambridge, Mass.: The Museum, 1941).

Lanning, John Tate. *Academic Culture in the Spanish Colonies* (London: Oxford University Press, 1940).

Leite Lopes, José. *La ciencia y el dilema de América Latina: dependencia o liberación* (Buenos Aires: Siglo Ventiuno Argentina Editores, 1972).

Lemos, Fernando Cerqueira. "Contribuição à história do Instituto Bacteriológico, 1892–1940," *Revista do Instituto Adolfo Lutz* 14 (1954) (Número especial).

Leonard, Irving A. *Don Carlos de Sigüenza y Góngora, a Mexican Savant of the Seventeenth Century* (Berkeley: University of California Press, Publications in History 18, 1929).

———. "A Great Savant of Colonial Peru; Don Pedro de Peralta," *Philological Quarterly* 12 (January 1933), 54–72.

———. "Science, Technology, and Hispanic America," *The Michigan Quarterly Review* 2 (1963), 237–245.

Lima, J.P. Carvalho. "Instituto Adolfo Lutz," *Revista do Instituto Adolfo Lutz* 1 (1941), 5–20.

Lins, Ivan. *História do positivismo no Brasil* (São Paulo: Companhia Editôra Nacional, Brasiliana 322, 1964).

Lipset, Seymour Martin. "Values, Education and Entrepreneurship," *in* Lipset, S.M. and Solari, Aldo, eds. *Elites in Latin America* (Oxford University Press, 1967).

Lobo, Francisco Bruno. *O ensino da medicina no Rio de Janeiro. Homeopatia.* (Rio de Janeiro: Oficina Gráfica da Universidade Federal do Rio de Janeiro, 1968).

———. *Uma universidade no Rio de Janeiro*, 2 vols. (Rio de Janeiro: Oficina Gráfica da Universidade Federal do Rio de Janeiro, 1967–69).

Lobo, Francisco Bruno, ed. *Memória histórica de 1864, redigida pelo professor João Joaquim de Gouvea, lente de physiologia* (Rio de Janeiro, 1958). Mimeographed.

———. ed. *Memórias históricas dos acontecimentos de 1855 c 1856 apresentadas à congregação dos lentes da Faculdade de Medicina do Rio de Janeiro em cumprimento do art. 197 dos estatutos.* Pelo Dr. Thomas Gomes dos Santos, lente de hygiene da mesma Faculdade (Rio de Janeiro: Typographia Universal de Laemmert, 1857). Mimeographed.

———. ed. *Memória histórica dos acontecimentos notaveis da Faculdade de Medicina do Rio de Janeiro, succedidos durante o anno de 1861*, apresentada à congregação em cumprimento do que determina o art. 197 dos estatutos, pelo Dr. Antônio Ferreira Pinto, lente substituto da secçao médica. Mimeographed.

———. ed. *Memória histórica dos accontecimentos mais notaveis occorridos na Faculdade de Medicina do Rio de Janeiro em 1879*, redigida pelo Dr. Nuno de Andrade, lente substituto da secção médica. Mimeographed.

———. ed. *Relatório do director da Faculdade de Medicina do Rio de Janeiro por 1883*, pelo Visconde de Saboa. Mimeographed.

———. ed. *Memórias históricas da Faculdade de Medicina e Pharmacia do Rio de Janeiro em 1894*, pelo Dr. José Maria Teixeira, lente de pharmacologia e arte de formular. Mimeographed.

———. ed. *O ensino da medicina no Rio de Janeiro*, 2 vols. (Rio de Janeiro: Oficina Gráfica da Universidade do Brasil, 1964).

Longstaff, George B. *Butterfly-Hunting in Many Lands; Notes of a Field Naturalist* (London: Longmans, Green and Co., 1912).

Lurie, Edward. *Louis Agassiz. A Life in Science* (Chicago: University of Chicago Press, 1960).

Lutz, Adolpho. "Contribuição à história da medicina no Brasil segundo os relatórios do Adolpho Lutz como director do Instituto Bacteriológico de São Paulo (1893–1908)," *Memorias do Instituto Oswaldo Cruz* 39 (1943), 177–252.

———. "Observações sôbre as molestias da cidade e do estado de São Paulo," *Revista Médica de São Paulo* 1 (1898), 4–6, 39–41, 60–61, 95–99.

———. "Relatório dos trabalhos do Instituto Bacteriológico durante o anno de 1897," *Revista Médica de São Paulo* 1 (1898), 175–187.

———. "Reminiscências da febre amarella no Estado de São Paulo," *Memorias do Instituto Oswaldo Cruz* 24 (1930), 127–142.

———. "Reminiscências da febre typhoide," *Memorias do Instituto Oswaldo Cruz* 31 (1936), 851–865.

———. "Resumo dos trabalhos do Instituto Bacteriológico de São Paulo, 1892 à 1906," *Revista Médica de São Paulo* 10 (1907), 65–87.

———. "Trabalhos do Instituto Bacteriológico do Estado de São Paulo durante o anno de 1898," *Revista Médica de São Paulo* 2 (1899), 300–321.

Lyra, Heitor. *História de Dom Pedro II, 1825–1891*, 3 vols. (Sao Paulo: Companhia Editôra Nacional, Brasiliana Vols. 133A, 133B, 1938–40).

MacGowan, Kenneth, and Hester, Joseph A., Jr. *Early Man in the New World* (Garden City, New York: Anchor Books, Doubleday and Company, Inc., 1962).

Magalhães, Fernando. *O centenário da Faculdade de Medicina do Rio de Janeiro, 1832–1932* (Rio de Janeiro: A. P. Barthel, 1932).

Magalhães, Octavio de. "Alcides Godoy," *Memorias do Instituto Oswaldo Cruz* 49 (1951), 1–6.

Magalhães, Ruth Martins de. "A biblioteca do Instituto Oswaldo Cruz." Lecture given at a conference held in the Department of Pathology, Oswaldo Cruz Institute, 18 May 1966. Mimeographed.

Manizer, G.G. *A expedição do acadêmico G. I. Langsdorff ao Brasil (1821–1828)* (São Paulo: Companhia Editôra Nacional, Brasiliana 329, 1967).

Manning, Thomas G. *Government in Science: The U.S. Geological Survey, 1867–1894* (Lexington: University of Kentucky Press, 1967).

Marchant, Anyda. "Dom João's Botanical Garden," *Hispanic American Historical Review* 41 (May 1961), 259–274.

———. *Viscount Mauá and the Empire of Brazil: A Biography of Ireneu Evangelista de Sousa, 1813–1889* (Berkeley: University of California Press, 1965).

Marchoux, S. and Simond, P.L. "A febre amarella. Relatório da Missão Francesa," *Revista Médica de São Paulo* 7 (1) (1904), 12–21, 38–42, 61–66.

Martin, Thomas. *The Royal Institution* (London: Longmans Green and Company, 1944).

Mattos, Anibal. *O sábio Dr. Lund e estudos sôbre a pre-história brasileira. Vida e obra de Peter Wilhelm Lund, ethnographia, archeologia, anthropologia, antiguidade do homem no Brasil* (Bello Horizonte: Edições Apóllo, 1935).

Mello Leitão, Candido F. de. *A biologia no Brasil* (São Paulo: Companhia Editôra Nacional, Biblioteca pedagógica brasileira. Ser. 5a, Brasiliana 99, 1937).

———. *História das expedições científicas no Brasil* (São Paulo: Companhia Editôra Nacional, Biblioteca pedagógica brasileira. Ser. 5a, Brasiliana 209, 1941).

Merton, Robert K. *Science, Technology and Society in Seventeenth Century England* (New York: Harper and Row, 1970).

Miller, Howard Smith. *Dollars for Research; Science and its Patrons in Nineteenth-Century America* (Seattle: University of Washington Press, 1970).

Miranda Ribeiro, Alipio de. *A Commissão Rondon e o Museu Nacional. Conferências* (Rio de Janeiro: L. Macedo, 1920).

Moacyr, Primitivo. *A instrução e a República*, 4 vols. (Rio de Janeiro: Imprensa Nacional, 1941–42).

———. *A instrução Pública no Estado de São Paulo*, 2 vols. (São Paulo: Companhia Editôra Nacional, Biblioteca pedagógica brasileira, Ser. 5a, Brasiliana 213, 213A, 1942).

Moll, Aristides A. *Aesculapius in Latin America* (Philadelphia and London: W.B. Saunders Company, 1944).

Moravcsik, Michael T. "Technical Assistance and Fundamental Research in Underdeveloped Countries," *Minerva* 2 (Winter 1964), 197–209.

———. *Science Development: Toward the Building of Science in Less Developed Countries* (Bloomington, Indiana: International Development Research Center, Indiana University, PASITRAN, 1975).

Moreira, Juliano. "Da necessidade da fundação de laboratórios nas hospitães," *Revista Médica de São Paulo* 5 (1902), 258–262.

———. "O progresso das sciências no Brasil. Conferência realisada à 24 de outubro de 1912," *Annaes da Bibliotheca Nacional do Rio de Janeiro* 35 (1913), 32–47.

Morse, Richard. *From Community to Metropolis: A Biography of São Paulo,*

Brazil (Gainesville, Florida: University of Florida Press, 1958).
Motten, Clement G. *Mexican Silver and the Enlightenment* (Philadelphia: University of Pennsylvania Press, 1950).
Müller, Fritz. *Facts and Arguments for Darwin.* With Additions by the Author. Translated from the German by W.S. Dallas (London: J. Murray, 1869).
Nader, Claire, and Zahlan, A.B., eds. *Science and Technology in Developing Countries* (Berkeley and Los Angeles: California University Press, 1967).
Nascimento, Alfredo. *O centenário da Academia Nacional de Medicina do Rio de Janeiro, 1829–1929.* (Rio de Janeiro: Imprensa Nacional, 1929).
Naylor, Bernard. *Accounts of Nineteenth-Century South America; An Annotated Checklist of Works by British and United States Observors* (London: The Athlone Press, University of London, 1969).
Nayudamma, Y. "Promoting the Industrial Application of Research in an Underdeveloped Country," *Minerva* 5 (Spring 1967), 323–339.
Neiva, Arthur. "Formação de raça do hemotozoario do impaludismo resiztente à quinina," *Memorias do Instituto Oswaldo Cruz* 2 (1910), 131–140.
———. *Esboço histórico sôbre a botânico e a zoologia no Brasil* (São Paulo: Impressora Paulista, 1926).
———. "Necrologia do Dr. Adolpho Lutz, 1855–1940," *Memorias do Instituto Oswaldo Cruz* 36 (1941), i–xxiii.
———. "Profilaxia da malaria e trabalhos de engenheira: notas, comentários, recordações," *Revista do Clube de Engenheria* 6 (1940), 60–75.
———. and Penna, Belisário. "Viagem científica pelo norte da Bahia, sudoeste de Pernambuco, sul de Piauhí, e de norte à sul de Goiáz," *Memorias do Instituto Oswaldo Cruz* 8 (1916), 74–224.
Nelson, Richard R. "'World Leadership,' the 'Technological Gap' and National Science Policy," *Minerva* 9 (July 1971), 386–399.
Netto, Americo R. "O caminho para a formação de serviço sanitário de São Paulo de 1597 à 1891," *Arquivos de Hygiene e Saúde Pública* 7 (Janeiro 1942), 5–34.
Oliveira, Octavio G. de. *Oswaldo Cruz e suas atividades na direção da saúde pública brasileira* (Rio de Janeiro: Serviço Gráfico do Instituto Brasileiro de Geografia e Estatística, 1955).
Orbigny, Alcide Dessalines d'. *Voyage dans l' Amérique Méridionale (le Brésil, la république orientale de L'Uruguay, la République argentine, la Patagonie, la république du Chili, la république de Bolivia, la république du Pérou)*, exécuté pendant les années 1826, 1827, 1828, 1829, 1830, 1831, 1832 et 1833 (Paris: Pitois-Levrault, 1835–47).
Paiva, L. Miller de. "História científica do Instituto Butantã," *Publicações Médicas* 17 (Maio-Julho 1946), 59–72.
Pan American Health Organization. Advisory Committee on Medical Research. *Science Policy in Latin America; Substance, Structures, and*

Processes (Washington, D.C.: Pan American Sanitary Bureau, 1966).
Pan American Sanitary Bureau. *The Pan American Sanitary Bureau; Half a Century of Health Activities, 1902–1954* by Miguel E. Bustamente (Washington, D.C.: Pan American Sanitary Bureau, Miscellaneous Publications no. 23, 1952).
———. *Scientific Societies and Institutions in Latin America* (Washington, D.C. Pan American Sanitary Bureau, 1940).
Pasteur, Louis. "Le budget de la science," *Revue des cours scientifiques* 5 (1867–1868), 137–139.
Peller, Sigusmund. "Walter Reed, Carlos Finlay, and their Predecessors around 1800," *Bulletin of the History of Medicne* 33 (1959), 195–211.
Penna, Belisário. *Oswaldo Cruz. Impressões de um discípulo* (Rio de Janeiro: Revista dos Tribunaes, 1922).
———. *O saneamento do Brasil* (Rio de Janeiro: Joacintho Ribeiro dos Santos, 1923).
Penick, James L., *et al.*, eds. *The Politics of American Science, 1939 to the Present* (Chicago: Rand McNally, 1965).
Peregrino, Umberto. *História e projeção das instituições culturais do exército* (Rio de Janeiro: José Olympio, 1967).
Pereira, José Verissimo da Costa. "Henri Gorceix," *Revista Brasileira de Geografia* 5 (1943), 627–630.
Perreira da Silvo, Gastão. *O romance de Oswaldo Cruz* (Rio de Janeiro: Brasília Editôra, n.d.)
Pestana, Bruno Rangel. "Cinquentenário do Instituto Adolfo Lutz," *Revista do Instituto Adolfo Lutz* 2 (1942), 181–190.
Pinto, César. "Arthur Neiva: cientista e homem público," *Revista Médica-Cirurgia Brasileira* 40 (1932), 2–10.
Poppino, Rollie E. *Brazil, the Land and People* (New York: Oxford University Press, 1963).
Price, Derek J. de Solla. *Little Science, Big Science* (New York: Columbia University Press, 1963).
———. "Is Technology Historically Independent of Science? A Study in Statistical Historiography," *Technology and Culture* 6 (1965), 553–568.
———. "Measuring the Size of Science," *Proceedings of the Israel Academy of Sciences and Humanities* 4 (1969), 98–111.
———. *The Relations Between Science and Technology and Their Implications for Policy Formation* (Stockholm, Sweden: FOA Reprints No. 26, 1972/73), pp. 16–17.
Price, Don K. *Government and Science; Their Dynamic Relation in American Democracy* (Oxford: Oxford University Press, 1962).
———. *The Scientific Estate* (Cambridge, Mass.: The Belknap Press of Harvard University Press, 1965).

Pursell, Carroll W., Jr., ed. *Readings in Technology and American Life* (New York: Oxford University Press, 1969).

Puech, L. Rezende. *Sociedade de Medicina e Cirurgia de São Paulo: Membria histbrica, 1895–1921* (São Paulo: Casa Garraux, 1921).

Quarto Congresso Médico Latino Americano, *A medicina no Brasil* (Rio de Janeiro: Imprensa Nacional, 1908).

Raeders, Georges. *Pedro II e os sábios franceses* (Rio de Janeiro: Atlântica Editôra, 1944).

Randall, John H. *The School of Padua and the Emergence of Modern Science* (Padova: Ed. Antenore, 1961).

Rathbun, Richard. *Sketch of the Life and Scientific Work of Professor C.F. Hartt.* Read Before the Boston Society of Natural History (Boston: 1878).

Ravetz, Jerome R. *Scientific Knowledge and its Social Problems* (Oxford: Clarendon Press, 1971).

Reed, Walter, *et al.* "The Etiology of Yellow Fever: A Preliminary Note," *American Public Health Association, Public Health Papers and Reports* 22 (1900), 37–53.

Reingold, Nathan. *Science in Nineteenth-Century America, A Documentary History* (New York: Hill and Wang, 1964).

Revista do Instituto Adolfo Lutz 15, 1955. (Número único. Número comemorativo do centenário do Nascimento do Adolfo Lutz).

Ribas, Emílio, *O mosquito como agente da propagação da febre amarella* (São Paulo: Diário Official 1901).

———. "Prophylaxia da febre amarella: memória apresentada ao 5° Congresso Brasileiro da Medicina e Cirurgia," *Revista Médica de São Paulo* 6 (1903), 477–485, and 504–516.

———. "Relatório apresentado ao Sr. Dr. Secretário dos Negocios Interiores e da Justiça (1905)," *Revista Médica de São Paulo* 9 (1906), 257–262.

———. "Relatório referente ao anno de 1906 apresentado pelo Dr. Emílio Ribas, Director do Serviço Sanitário ao Sr. Sec. dos Negocios do Interior," *Revista Médica de São Paulo* 10 (1907), 213–236.

———. "A extinção da febre amarella no Estado de São Paulo (Brasil) e na cidade do Rio de Janeiro," *Revista Médica de São Paulo* 12 (1909), 198–209.

———. "Communicações. Quinto Congresso Brasileiro da Medicina e Cirurgia, 1903. (Rio de Janeiro, 16 de junho à 2 de julho de 1903). O mosquito e a febre amarella, trabalho da Directoria do Serviço Sanitário de São Paulo,' *Archivos de Hygiene e Saúde Pública 1–2* (1936), número especial, 270–306.

Ribeiro, Leonídio. *Medicina no Brasil* (Rio de Janeiro:Imprensa Nacional, 1940).

Ribeiro, Lourival. *Medicina no Brasil colonial* (Rio de Janeiro: Editorial Sul Americana, 1971).

Rippy, J. Fred. *Latin America and the Industrial Age* (New York: G.P. Putnam's Sons, 1944).

Rio de Janeiro. Museu Nacional, *Guia da exposição anthropolbgica brasileira realizada pelo Museu Nacional do Rio de Janeiro, à 29 de julho de 1882* (Rio de Janeiro: G. Leuzinger e Filhos, 1882).

———. ———. *João Batista de Lacerda. Comemoração do centenário de nascimento, 1846–1946* (Rio de Janeiro: Museu Nacional, Publicações avulsas no. 6, 1951).

———. ———. *José Otiticica Filho, As publicações do Museu Nacional como contribuição para a ciência e a cultura* (Rio de Janeiro: Oficina Gráfica da Universidade do Brasil, Publicações avulsas do Museu Nacional n. 42, 1961).

Rocha Lima, Henrique de. "Com Oswaldo Cruz em Manguinhos," *Ciência e Cultura* IV, No. 1 e 2, 15–21.

Roche, Marcel. "Social Aspects of Science in a Developing Country," *Impact of Science on Society* 16 (1966), 51–60.

———. "Science in Spanish and Spanish American Civilization," *in* Ciba Foundation Symposium 1 (new series), *Civilization and Science: In Conflict or Collaboration?* (Amsterdam: Elsevier, 1973).

Rockefeller Foundation. *Annual Report*, 1919, 1921.

———. *Directory of Fellowship Awards, 1917–1950* (New York, N.Y.).

Roquette, Paulo. "O Museu Nacional e a educação brasileira," *Jornal do Comercio*, July 10, 11, 1933.

Rosenberg, Charles E. *The Cholera Years, The United States in 1832, 1849, and 1866* (Chicago; University of Chicago Press, 1962).

Ross, Ronald. *The Prevention of Malaria* (London: J. Murray, 1910).

Rossi, Peter H. "Researchers, Scholars and Policy Makers: The Politics of Large Scale Research," *Daedalus* 93 (Fall 1964), 1142–1161.

Ruttan, V. W., and Hayami, Yujrio. "Technology Transfer and Agricultural Development," *Technology and Culture* 14 (1973), 119–151.

Sagasti, Francisco R. "Discussion Paper: Underdevelopment, Science and Technology: The Point of View of the Underdeveloped Countries," *Science Studies* 3 (1973) 47–59.

Sahagún, Bernadino de. *Historia general de las cosas de Nueva España, que en doce libros y dos volumunes escribib, el R.P. fr. Bernadino de Sahagún.* Dala a luz connotas y suplementos Carlos María de Bustamente (México: Imp. del ciudadano A. Valdés, 1829–30).

Sallos, Pedro. *Histbria da medicina no Brasil* (Belo Horizonte: Editôra G. Holman Ltda, 1971)

Salomon, Jean-Jacques. *Science and Politics* (Cambridge, Mass: The M.I.T. Press, 1973).

Santos Filho, Lycurgo. *Histbria da medicina no Brasil: do século XVI ao século*

XIX (São Paulo: Editôra Brasiliense Ltda., 1947).

Sayers, Raymond S., ed. *Portugal and Brazil in Transition* (Minneapolis: University of Minnesota Press, 1968).

Schofield, Robert E. *The Lunar Society of Birmingham; A Social History of Provincial Science and Industry in Eighteenth Century England* (Oxford: Clarendon Press, 1963).

Schmookler, Jacob. *Invention and Economic Growth* (Cambridge, Mass.: Harvard University Press, 1966).

Scott, H. Harold. *A History of Tropical Medicine,* 2 vols. (Baltimore: The Williams and Wilkins Company, 1942).

Senn, Nicholas. "Notas de viagem do América do Sul," *Revista Médica de São Paulo* 11 (1908), 107-108.

Serpa, Phoción. *Osvaldo Cruz, el Pasteur del Brasil, vencedor de la fiebre amarilla* (Buenos Aires: Editorial Claridad, 1945).

Shaplen, Robert. *Toward the Wellbeing of Mankind: Fifty Years of the Rockefeller Foundation.* Foreword by J. George Harrar. Edited by Arthur Bernon Tourtellot (Garden City, New York: Doubleday, 1964).

Shils, Edward. *The Criteria for Scientific Development: Public Policy and National Goals; A Selection of Articles from Minerva* (Cambridge, Massachusetts: The M.I.T. Press, 1968).

Shryock, Richard Harrison. *Medicine in America, Historical Essays* (Baltimore, Maryland: The Johns Hopkins Press, 1966).

Sigaud, Joseph F.X. *Du climat et des maladies du Brésil; ou Statistique médicale de cet empire* (Paris: Fortin, Masson, 1844).

Silva, Pedro Dias de, *et al.* "Notas para a memória histórica da Faculdade de Medicina de São Paulo," *Annaes da Faculdade de Medicina de São Paulo* 1 (1926), 1-77.

Silvert, Kalman, ed. *The Social Reality of Scientific Myth, Science and Change* (New York: American Universities Field Staff, Inc., 1969).

Singer, Charles, and Underwood, E. Ashworth, *A Short History of Medicine* (New York and Oxford: Oxford University Press, Second Edition, 1962).

Smallwood, William M., and Smallwood, Mabel S. C. *Natural History and the American Mind* (New York: Columbia University Press, 1941).

Sociedade Médica e Cirurgica de São Paulo, "Febres paulistas, paracer da Sociedade e conclusões finaes, 1 e 13 dezembro," *Boletim da Sociedade Médica e Cirurgica de São Paulo* 3 (1897).

Sodré, Nelson Werneck. *Panorama do segundo império* (São Paulo: Companhia Editôra Nacional, Biblioteca pedagógica brasileira, Ser. 5a, Brasiliana 170, 1939).

Souza, Heitor G. de, *et al. Política científica* (São Paulo: Editôra Perspectiva, 1972).

Souza Campos, Ernesto de. *Educação superior no Brasil* (Rio de Janeiro: Serviço Gráfico do Ministerio Superior da Educação, 1940).

———. *Instituições culturais e de educação superior no Brasil: resumo histórico* (Rio de Janeiro: Imprensa Nacional, 1941).

Souza Reis, Alvaro A. de. *História da literatura médica brasileira* (Rio de Janeiro: Livraria J. Leite, n.d.).

Stearns, Raymond P. *Science in the British Colonies of America* (Urbana: University of Illinois Press, 1970).

Stein, Stanley. *Vassouras. A Brazilian Coffee County, 1850–1890* (New York: Atheneum, 1970).

Stein, Stanley J., and Stein, Barbara H. *The Colonial Heritage of Latin America; Essays on Economic Dependence in Perspective* (New York: Oxford University Press, 1970).

Stocking, George W., Jr. *Race, Culture and Evolution: Essays in the History of Anthropology* (New York: The Free Press, 1968).

Struik, Dirk J. *Yankee Science in the Making* (Boston: Little, Brown, 1948).

Sunkel, Oswaldo. "Underdevelopment, the Transfer of Science and Technology and the Latin American University," *Human Relations* 24 (1971), 1–18.

Thorndike, Lynn. *A History of Magic and Experimental Science*, 8 vols. (New York: Columbia University Press, 1923–58).

Torres, Théophilo. *La campagne sanitaire au Brésil: faits et documents* (Paris: Societé Generale d'Impression, 1913).

Torres Homem, João Vicente. *Estudo clínico sôbre as febres do Rio de Janeiro* (Rio de Janeiro: Lopes do Couto e Cie, Editores, 1886).

Townes, Charles H. "Quantum Electronics and Surprise in the Development of Technology, The Problem of Research Planning," *Science* 159 (1968), 699–703.

Tsuge, Hideomi, ed. *Historical Development of Science and Technology in Japan* (Tokyo: Kokusai Bunka Shinkokai, The Society for International Cultural Relations, 1961).

UNESCO. *Principles and Problems of National Science Policies.* Meeting of the Co-ordinators of Science Policy Studies, Karlovy Vary (Czechoslovakia), 6–11 June 1966 (Paris: UNESCO, 1967).

———. *Structural and Operational Schemes of National Science Policy.* Science Policy and Research Organization in the Countries of North Africa and the Middle East. Algiers, 20–26 September 1966 (Paris: UNESCO, 1967).

United Nations. *Scientific Institutions and Scientists in Latin America, Instituciones científicas y científicos latinoamericanos* (Montevideo: Centro de Cooperación Científica Para América Latina, 1949–64).

———. *Education, Human Resources and Development in Latin America* (United Nations, New York: Economic Commission for Latin America, 1968).

United States Treasury Department. Marine Hospital Service, *Report on the Etiology and Prevention of Yellow Fever.* By George M. Sternberg

(Washington: Government Printing Office, 1890).

Vaitsos, Constantino V. *Comercializacibn de tecnologla en el Pacto Andino* (Lima, Peru: Instituto de Estudios Peruanos, 1973).

———. "Power, Knowledge and Development Policy: Relations Between Transnational Enterprises and Developing Countries," (Uppsala, Sweden: The Dag Hammarskjöld Foundation, 1974), unpublished manuscript.

Vallery-Radot, Pasteur. "Les Instituts Pasteur d'outre mer," *La Presse Mèdicale* 21 (Mars 1939), 410–413.

———. ed. *Oeuvres de Pasteur* 7 vols. (Paris: Masson Oce., Editeurs, 1922–39).

Van Tassel, David D., and Hall, Michael G., eds. *Science and Society in the United States* (Homewood, Illinois: The Dorsey Press, 1966).

Vauthier, Louis Lèger. *Diàrio ìntimo do engenheiro Vauthier, 1840–1846* (Rio de Janeiro: Serviço Gràfico do Ministèrio da Educação e Saùde, 1940).

Vaz, Eduardo. *Fundamentos da histbria do Instituto Butantã (São Paulo: Instituto Butantã, 1949).*

———. "Vital Brasil," *Anais Paulistas de Medicina e Cirurgia* 60 (1950), 334–366.

———. *Vital Brasil* (São Paulo: São Paulo Editôra, 1950).

Veysey, Laurence R. *The Emergence of the American University* (Chicago: University of Chicago Press, Phoenix Books, 1970).

Vianna, Gaspar de Oliveira. "Contribuição para o estudo da anatomia patalòjica da 'Molestia de Carlos Chagas,' (Esquizotrypanoze humana ou tireoidite parazitària)," *Memorias do Instituto Oswaldo Cruz* 3 (1911) 276–294.

Vieira, Francisco Borges. "Campinas e a obra de Emìlio Ribas," *Arquivos de Hygiene e Saùde Publica* 5 (1940), 95–111.

———. "Primeiros tempos da administração sanitària paulista e seus antecedentes no paìs," *Arquivos de Hygiene e Saùde Pùblica* 8 (1943), 33–44.

Vieira Filho, J. "Antônio Cardosa Fontes e sua obra," *Jornal de Comércio*, 28 de dezembro, 1932.

Von Hägen, Victor Wolfgang. *South America Called Them; Explorations of the Great Naturalists: La Condamine, Humboldt, Darwin, Spruce* (New York: Alfred A. Knopf, 1945).

———. *The Green World of the Naturalists: A Treasury of Five Centuries of Natural History in South America* (New York: Greenberg Publishers, 1948).

Weinberg, Alvin M. "Scientific Choice and Biomedical Science," *Minerva* 4 (Autumn 1965), 3–14.

———. *Reflections on Big Science* (Cambridge, Mass.: The M.I.T. Press, 1967).

Westfall, Richard S. *Science and and Religion in Seventeenth-Century England* (New Haven: Yale University Press, 1958).

Whitaker, Arthur P. *Latin America and the Enlightenment* (New York: Cornell University Press, 2nd ed., Great Seal Books, 1961).

Wightman, W.P.D. *Science and the Renaissance* (Edinburgh: Oliver and Boyd, 1962).

Wilkins, W.H. *The Romance of Isabel Lady Burton*, 2 vols. (New York: Dodd Mead and Co., 1897).

Williams, Mary Wilhelmine. *Dom Pedro the Magnanimous, Second Emperor of Brazil* (Chapel Hill: University of North Carolina Press, 1937).

Wilson, Iris Higbie. "Scientists in New Spain: The Eighteenth Century Expeditions," *Journal of the West* 1 (July/October 1962), 24–44.

Wionczek, Miguel S. *Inversión y tecnología extranjera en América Latina* (México: Editorial de Joaquín Mortiz, 1971).

World Health Organization. Technical Report Series, No. 202, *Chagas' Disease. Report of a Study Group* (Geneva: World Health Organization, 1960).

———. ———, No. 411, *Comparative Studies of American and African Trypanosomiasis. Report of a WHO Scientific Group* (Geneva: World Health Organization, 1969).

Plates

Photographs reproduced with the permission of the Oswaldo Cruz Institute.

I Oswaldo Cruz in 1903, when director of the Serum Therapy Institute in Rio de Janeiro and of the campaign against yellow fever, bubonic plague, and smallpox.

II The nearly completed buildings of the Oswaldo Cruz Institute in 1908, showing the remains of the original laboratory at Manguinhos on the right, and the bay of Guanabara beyond.

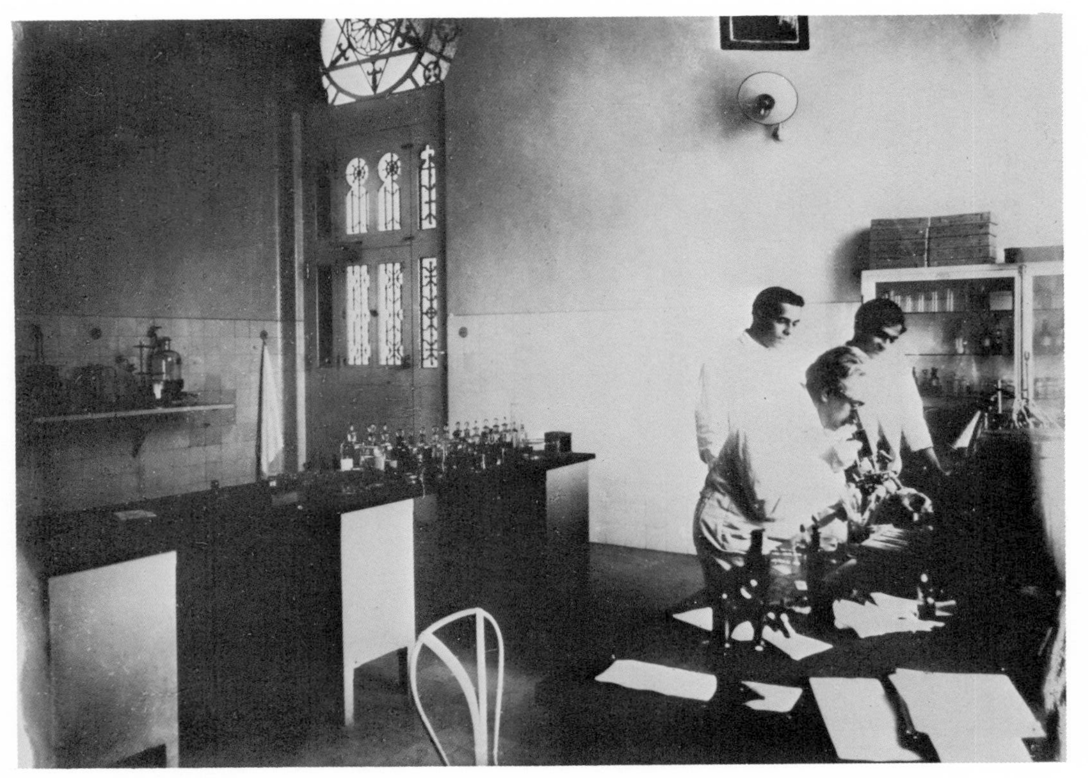

III Oswaldo Cruz in one of the laboratories of the Oswaldo Cruz Institute.

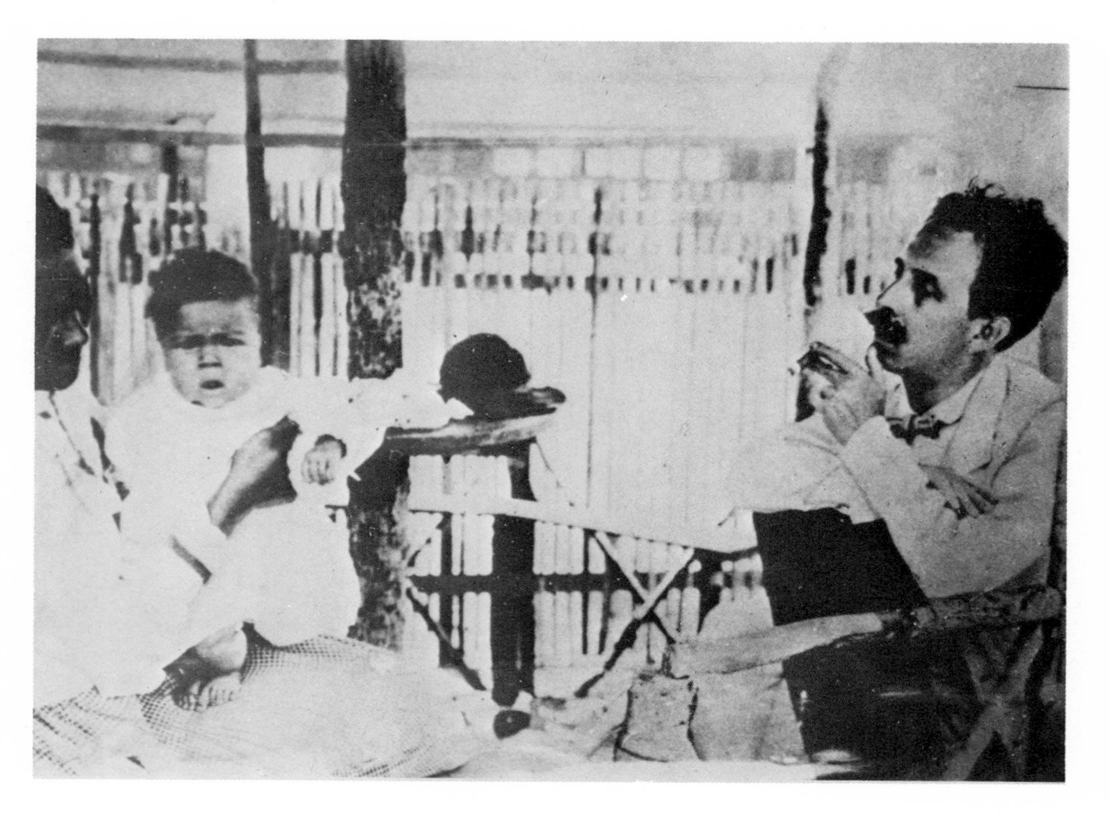

IV Carlos Chagas, discoverer of American sleeping sickness (*Trypanosomiasis americana*), examining a sick child in Lassance, Minas Gerais, in 1909.

V Staff and visiting scientists in front of the dining room at the Oswaldo Cruz Institute in 1908. *Seated, from left to right:* Carlos Chagas, José Gomes de Faria, Antônio Cardoso Fontes, Max Hartmann, Oswaldo Cruz, Stanislaus von Prowazek, Adolfo Lutz. *Standing, from left to right:* Artur Neiva, Henrique da Rocha Lima, Henrique Figueiredo de Vasconcellos, Henrique de Beaurepaire Aragão, Alcides Godoy.

VI A newspaper cartoon published during the public health campaign in Rio de Janeiro, showing the supposed horrors of compulsory vaccination against smallpox.

VII By 1908 the attitude toward Oswaldo Cruz has changed. Now a newspaper cartoon portrays Cruz as a hero, and bears the caption: "This is the way the country receives its sons that honor and love the nation; it crowns and blesses them."

VIII A cartoon in French, showing the public's view of the dominating presence of Oswaldo Cruz at the Oswaldo Cruz Institute and the kind of scientific work carried out there.

Index

Affonso, Baron Pedro, 69, 139; director of Federal Vaccination Institute, 68; advises municipality of Rio on plague, 68; appointed director of Serum Therapy Institute, 68; appoints staff, 68–69; resigns from Serum Therapy Institute, 75
Agassiz, Louis: with Thayer expedition, 27, 30; publication of *A Journey to Brazil*, 30; comments on Brazilian scientific institutions, 30–31; his effect on Brazilian science, 30; career in United States, 35; as entrepreneur, 36
Alves, Francisco Rodrigues, President of Brazil, 57, 89, 90, 91, 92, 96, 97, 144, 146; as governor of São Paulo, 84–85; need to renovate Rio de Janeiro, 85; and yellow fever campaign, 86
American sleeping sickness, *see* Trypanosomiasis americana
Andrada, José Bonifácio de, 22
Andrade, Nuno de, 33; report on Medical School of Rio de Janeiro, 54; as director of public health in 1902, 85; publication on Finlay doctrine, 90
Applied science: importance at Oswaldo Cruz Institute, 112–115; need to be balanced with basic science, 115; overemphasis in Bacteriological Institute of São Paulo, 145, 151; need to be integrated with basic research in developing countries, 164–165
Aragão, Henrique de Beaurepaire, 111, 121, 128; work on *plasmodium* of pigeon, 98, 118; joins Oswaldo Cruz Institute, 107; as staff member of Institute, 109
Araújo, Silva, 54
Azevedo, Fernando de: comment on Oswaldo Cruz Institute, 125

Bacon, Francis, 160
Bacteriological Institute of São Paulo: founded, 9; failure to expand, 9, 144–150; Martin Ficker at, 111, 150–152; synopsis of history, 135–136; scientific work, 138–144; budget, 139; difficulty in staffing, 139; Adolfo Lutz appointed director, 139; investigation of cholera, 140–141; example of applied science trap, 145–147, 165; poor facilities, 147–150; failure to recruit students, 147; heavy work load, 149, 150–152
Barbosa, Rui, 29
Barnuevo, Pedro de Peralta, 17
Barreto Burgos, Cariolano, 147
Barreto, Luís Pereira, 56; letter to Ribas on yellow fever, 58, 142–143; and yellow fever, 59, 84; inoculation experiments, 84–85
Basalla, George, 79; model of spread of western science, 14; definition of "colonial" science, 14, 16, 20, 36; definition of "stage three" independent science, 15; criticism of his model, 15, 19, 37; application of model to Spanish America, 16
Bates, Henry W., 27
Ben-David, Joseph, 158; on rise of research science, 39–40; on rise of research institutions, 99; research ideal attached to applied sciences, 122; on bacteriology, 122; need to cultivate research, 151
Biological Institute of São Paulo, *see* Instituto Biológico
Bonifácio, José, 94
Boxer, Charles R., 21, 22
Brasil, Vital, 142, 147, 148; work on plague in Santos, 66; appointed to Instituto Butantã, 67
Brazilian science: comparison with Spanish America, 20–21; in the seventeenth and eighteenth centuries, 22–23; and printing presses in eighteenth century, 22; influence of Enlightenment, 22, 23; effect of expulsion of Jesuits, 22; growth in the nineteenth century, 23–36; effect of transfer of court to Brazil, 23–25; founding of medical schools in Rio and Bahia, 24; Military Academy founded, 25; natural science in the nineteenth century, 25; role of exploration by foreigners, 26–28; Royal Botanical Garden, 27; Imperial Museum (later National Museum), 27;

new institutions in nineteenth century, 24–26, 27; in the period of the Old Republic, 29–40; role of foreigners in, 29–32, 33, 34–36; geology in nineteenth century, 34; factors in transformation of colonial science, 36–40; absence of private patronage, 38, 77; situation in science by 1900, 36–40; effect of rise of research science on, 38–40; effect of Darwinism, 56; effect of positivism, 56; effect of universities on, 127
Brooks, Harvey, 162
Bubonic plague: arrival in Brazil, 65, 66, 142; effect on Brazilian science, 65, 142; first cases in Rio, 67; deaths from in 1900 in Rio, 67; investigated by Lutz and Brasil, 142
Burton, Richard: medical treatment in Brazil, 51–52
Butantã Institute, 79, 147; founded, 67; later history, 126

Cabral, Pedro Alvares, 21
Carey, Elizabeth, 35
Carlos III, King of Spain, 19
Carvalho, Vincente de, 148
Castro, Francisco de, 70
Central Leather Research Institute of India, 180
Chagas, Carlos Ribeiro Justiniano, 98, 112, 113, 123, 127, 128; discovery of Trypanosomiasis americana, 6, 118–119; succeeds Cruz as director of Oswaldo Cruz Institute, 6, 124; early training, 109; recruitment to Oswaldo Cruz Institute, 109; debate on identity and nature of Trypanosomiasis americana, 119; influenza epidemic of 1918, 126; appointed director of public health, 126–127
Chagas' disease, *see* Chagas, Carlos Ribeiro Justiniano *and* Trypanosomiasis americana
Chamberland, Charles, 71
Chapot-Prévost, Eduardo: sent to Santos to diagnose cases of plague, 66, 67
Cholera: and public health legislation in Europe, 53; in state of São Paulo, 140; investigated by Lutz, 140–141
Cobo, Bernabé, 17
Condamine, Charles-Marie de la, 18
Copernicus, Nicholas, 16
Couty, Louis: appointed to Imperial Museum, 32; work with Lacerda, 32; analysis of Brazilian science, 56
Cruz, Bento Gonçalves, 69
Cruz, Oswaldo Gonçalves, 9, 47, 69, 73, 74, 79, 89, 90, 91, 96, 97, 107, 112, 113, 114, 119, 121, 128, 142, 144, 148; yellow fever campaign, 5, 87, 88; Oswaldo Cruz Institute named after, 6; investigation of bubonic plague in Santos, 67; appointed to Federal Serum Therapy Institute, 69; studies at Pasteur Institute, 70–72; appointed director of Serum Therapy Institute, 75; proposal for the Serum Therapy Institute in 1903, 75, 92; proposal for Serum Therapy Institute rejected, 77, 94; appointed director of public health, 87, 88; as administrator, 99; as entrepreneur, 99, 149; resigns from post of director of public health, 100; reports on sanitation conditions of the Amazon, 114; death, 124; political prominence, 146
Cunha, Aristides da, 118

Da Cunha, Euclides, 7
Daniels, George H.: on natural history in the United States in the nineteenth century, 26
Darwin, Charles, 27
Dependency: and science, 3–4; and technology, 167–169; bibliography, Ch. 8, ref. 25
Derby, Orville A.: article on Brazilian science in 1883, 30; brought to Brazil by Frederick Hartt, 31; appointed to Imperial Museum, 32; career in Brazil, 34–36; contributions to geology in Brazil, 34–35; geological work in São Paulo, 35; lack of influence in Brazil, 35–36
Dias, Ezequiel Caetano, 73; joins Oswaldo Cruz Institute, 107; as staff member of Institute, 109; leaves Oswaldo Cruz Institute, 111
Duclaux, Emile, 71, 72
Dupree, A. Hunter: on Texas cattle fever research, 121
Dutch science in Brazil, 21

Enlightenment, 18; in Latin America, 19
Ezequiel Dias Institute, 114

Faria, José Gomes de, 118
Faria, Rocha, 69
Federal Vaccination Institute, *see* Instituto Vaccínico do Distrito Federal
Ficker, Martin: comes to Brazil, 111; work at Bacteriological Institute of São Paulo, 145; critical report on Bacteriological Institute, 150; comments on role of research in hygiene sciences, 151–152; comments on Oswaldo Cruz Institute, 152
Finlay, Carlos, 90; theory of transmission of yellow fever, 59, 143
Flexner Report, 55
Fonseca, Emilia, 70
Fonseca, Olympio da, 97, 120, 123; rejection of Noguchi's work on yellow fever, 116
Fontes, Antônio Cardoso, 73, 107, 116, 128; as staff member of Oswaldo Cruz Institute, 109
Foreign explorations in Spanish America, 18
Foreign scientists: role in science in developing countries, 29–32, 33, 34–36, 78
Franklin, Benjamin, 14
Freire, Domingos, 29

Gazeta Médica da Bahia, 54
Godoy, Alcides: joins Oswaldo Cruz Institute, 108; work on *mal de ano*, 108; discovery of vaccine, 108; as staff member of Institute, 109
Goeldi, Emile, 32
Gomes de Faria, José, 116
Guyon, Félix, 73

Haffkine, Waldemar M., 66
Hartmann, Max, 118
Hartt, C. Frederick, 34; work in Brazil with the Imperial Geological Commission, 31; his *The Geology and Physical Geography of Brazil*, 31
Hirschman, Albert O., 168
Historical and Geographical Institute of Brazil, *see* Instituto Histórico e Geographico Brasileiro
Humboldt, Alexander von, 19, 26

Ihering, Hermann von, 33
Imperial Museum, *see* National Museum
Import substitution, industrialization and technology, 168
Institute of Experimental Pathology of Manguinhos, *see* Oswaldo Cruz Institute
Institutions of science: problems of leadership, 124; role of, 177
Instituto Bacteriológico de São Paulo, *see* Bacteriological Institute of São Paulo
Instituto Biológico, 126
Instituto Butantã, *see* Butantã Institute
Instituto Histórico e Geographico Brasileiro, 24, 27; exploration, 27, 28
Instituto Oswaldo Cruz, *see* Oswaldo Cruz Institute
Instituto Sôrothérapico Federal de Manguinhos, *see* Serum Therapy Institute of Manguinhos
Instituto Vaccínico do Distrito Federal, 68

João VI, King of Portugal and Brazil, 23, 24
Jungle yellow fever, 127

Kitasato, Shibasaburo, 65–66
Koch, Robert, 77

Lacerda, João Batista de, 39, 148; joins Imperial Museum, 32; publication of *Archivos do Museu Nacional*, 32; work on "bacillus" of yellow fever, 33
Landa, Diego de, 17
Langsdorff, Georg H. von, 26
Lanning, John Tate: on academic learning in Spanish America, 17–18
Latin America: lack of science in, 13; lack of discussion of science in, 13–14; *see* Brazilian science; *see* Spanish American science
Le Dantec, Félix: recruited to Bacteriological Institute of São Paulo, 139
Leinz, Victor: comment on Derby's work in geology, 35
Lemos, Fernando Cerqueira, 139
Leonard, Irving A., 17
Linnaeus, Carl, 19
Lisbôa, Henrique Marques, 107
Lisbôa, José da Silva, 24
Literary Society of Rio de Janeiro, *see* Sociedade Litterária do Rio de Janeiro
López, Hipólito Ruiz, 19
Lund, Peter Wilhelm, 28
Lutz, Adolfo, 9, 36, 67, 84, 112, 120, 123, 135, 142, 143, 148; work on

plague in Santos, 66; joins Oswaldo Cruz Institute, 111–112; biography, 139–140; appointed director of Bacteriological Institute of São Paulo, 139; work on leprosy, 140; study of cholera in São Paulo, 140; work on "Paulista" fevers, 141–142; work on malaria, 141; investigations into yellow fever in São Paulo, 142; yellow fever inoculation experiments, 144; leaves Bacteriological Institute, 144; problems as administrator, 148–149

Machado, Octávio, 107
Magalhães, Fernando, 107
Malaria: in Brazil, 113; and Oswaldo Cruz Institute, 113, 118; in state of São Paulo, 141; investigated by Lutz, 141
Malaspina, Alejandro, 19
Manguinhos, *see* Oswaldo Cruz Institute
Marcgraff, Georg, 21
Martius, Karl F.P., 26
Mattos, Deputy Mello, 93, 94, 97
Maurice, Prince of Nassau, 21
Maximilian of Wied-Neuwied, 26
Medical and Surgical Society of São Paulo: debate on "Paulista fevers," 141
Medical School of Bahia: founded, 24
Medical School of Rio de Janeiro, 96; founded, 24, 33; student population, 50–51; annual reports, 50; problems in nineteenth century, 54
Medicine in Brazil: founding of Rio and Bahia medical schools, 24, 56; medical tradition in the colonial period, 47–50; influence of epidemic disease on, 48–49; annual reports of medical school of Rio, 50–51, 54; social status of physicians, 51; medical journals, 51; public health legislation, 52; medical research in the nineteenth century, 53–54; problems and reform of medical schools, 53–56; effect of yellow fever on, 59
Memorias do Instituto Oswaldo Cruz: founded, 6; publications in, 116, 117
Mendonça, Artur Vieira de, 147
Mestro, João, 21
Metchnikoff, Elie, 71
Meyer, Carlos Luis: appointed interim director of the Bacteriological Institute of São Paulo, 147
Microbiology: course started at Oswaldo Cruz Institute, 110; at Oswaldo Cruz Institute, 120; "problem-oriented" field, 122
Microscopes: use of in Brazil, 55
Military Academy in Rio de Janeiro, 25, 90; reform of, 32; evaluated by Agassiz, 31
Mining School: founded in Ouro Prêto, 32
Miquel, Antonin Pierre, 97
Morães, Luis de, 96
Moreira, Juliano, 111
Morse, Richard M., 137, 138
Müller, Fritz, 28, 32
Multinationals: effect on integrated system of science and technology in developing countries, 168–170

Nader, Claire, 100, 170
National Museum (Museu Nacional do Rio de Janeiro), 33; founded as Imperial Museum, 27; history in nineteenth century, 27; reforms in nineteenth century, 33; stagnation, 78
"National" science: discussed, 120–124; and "international" science, 174; in developing country, 175
Natural history: in the United States, 25–26; in Brazil, *see* Brazilian science
Nayudamma, Y., 180
Neiva, Artur, 98, 113, 119, 121, 147; joins Oswaldo Cruz Institute, 108; as staff member of Institute, 109; travels abroad, 111; medical journey in Brazil, 115; work in public health in São Paulo, 126
Netto, Ladisláu de Souza Mello, 33, 39; on opportunities for a career in science in Brazil, 31–32; appointed to Imperial Museum, 32; criticisms of Imperial Museum, 32
Nobel prizes: in Latin America, 8
Noguchi, Hideyo: supposed discovery of cause of yellow fever, 116

Orbigny, Alcide Dessalines d', 26
Oswaldo Cruz Institute, 5, 48, 68, 96, 121, 139, 144, 160, 164, 165, 177, 179; yellow fever campaign in Brazil, 5; publication of *Memorias do*

Instituto Oswaldo Cruz, 6, 98; expansion of, 6; and model of Pasteur Institute, 75–79; effect of yellow fever campaign, 91–98; Cruz' plans for in 1903, 92; Congressional debate 1903, [illegible]–95; founded as the Institute of Experimental Pathology 1907, 97; budget tripled in 1907, 97; named after Cruz, 99; allowed to sell vaccines and serums, 99; analysis of reasons for success, 105–106, 134–135, 158–159; recruitment and training of medical staff, 106–112; development of library facilities, 112; weekly seminars at, 112; role of clients in survival, 112–115; role of applied science, 112–115; anti-malaria campaigns, 113, 118; yellow fever campaigns, 113–114; significance of public health work, 114; affiliated institute in Maranhão, 114; relation to expedition of General Rondon, 114; and Rockefeller Foundation, 115; importance of balance between basic and applied science, 115–120, 151–155; growth of numbers of researchers and publications, 116, 117; bacteriology at, 116; yellow fever research, 116; protozoological research at, 118–119; and discovery of Trypanosomiasis americana, 118–119; meaning of "national" science, 120, 173–174; significance of research for survival of, 123; expansion of activities, 123–124; weathers death of Cruz, 124; expansion of program under Chagas, 124; significance in Brazilian history, 5–7; 125–127; limitations of, 127–128
Overmeer, Hippólito Assueros: librarian of Oswaldo Cruz Institute, 112

Palestra Científica, 27
Pasteur Institute of Paris, 96; Cruz studies at, 70–72; founded, 71; organization, 71–72; as model for Oswaldo Cruz Institute, 71, 75–79; serum therapy at, 72; scientific missions abroad, 77–78
Pasteur, Louis, 54, 70, 134
Patterson, John L.: diagnosis of yellow fever in 1849, 54
Pavón, José, 19
Pedro II, Emperor of Brazil, 26; and science, 30; ties with France, 70; relation to Pasteur, 70–71
Pena, Afonso, 91, 97
Penna, Belisário: medical journey in Brazil, 115
Picanço, José Correi[illegible], 24
Piso, Wilhelm, 21
Polanyi, Michael: "the Republic of Science", 171
Politics of scientific choice: and applied science, 172; significance in developing country, 172–175
Polytechnical School in São Paulo, 135
Polytechnical School in Rio de Janeiro, 32–34
Pombal, Marquis of: and expulsion of Jesuits from Brazil, 22
Portugal: attitude to Brazil, 21–22; reform of University of Coimbra, 22
Powell, John Wesley, 34
Prowazek, Stanislaus von, 118
Public health in Brazil: public health legislation, 52; federal and municipal authority for, 86–87; new legislation proposed 1903, 88–89

Reed Commission, 84, 144; announcement of transmission of yellow fever by mosquito, 65
Research institutes: growth in industrialized world, 179; role in developing countries, 179–180
Research science: absence in Latin America, 13; absence in Brazil before 1900, 28–29, 33, 53; Agassiz comments on absence in Brazil, 30; Ben-David on, 99, 177; debate about need for in developing countries, 162–166
Ribas, Emílio, 36, 58, 84, 142; director of sanitary services in São Paulo, 140, 148; eradication of yellow fever in interior of São Paulo, 143
Rocha Lima, Henrique da, 98, 145, 150; recruitment to Oswaldo Cruz Institute, 108–109; professional training in Germany, 108–109; as staff member of Institute, 109; leaves Oswaldo Cruz Institute, 111; identification of lesions of yellow fever, 116; returns to Brazil, 126; work at Biological Institute of São Paulo, 126

Rockefeller Foundation: work in Brazil, 115, 127
Rodrigues Ferreira, Alexandre, 22
Rondon, General, 114
Rosa, João Ferreira da, 48
Roux, Emile, 69, 71; his "cours de microbie technique" at Pasteur Institute, 72, 110
Royal Scientific Expedition: to New Spain, 19
Royal Society of London, 16, 18

Sahagún, Bernardino de, 17
Sales Guerra, E., 70, 71, 87
Salomon, Jean-Jacques, 171; on science in industrialized countries, 161
Santos, Augusto Ferreira dos, 70
São Paulo, city: population growth, 136; Society for Propagating Public Instruction, 136–137; Law school, 136; rise of engineers and physicians after 1889, 137; yellow fever in, 138; reduction in mortality, 140
São Paulo Medical Society: debate on causes of yellow fever, 84
São Paulo, state: rise of science in, 36; arrival of bubonic plague, 65; formation of public health services, 66, 137–138; malaria in, 141; "Paulista" fevers in, 141; new US AID contract, 180
Science in developing countries: neglected by historians, 1–2; criteria of success, 8; definition of "success", 157; need for integrating basic and applied science, 159–169
Science in industrial world: the system of research and development, 159–162; founding of industrial laboratories, 160; effect of World War II, 160; integration of research, applied science, technology, 169–170
Science policy: implications of Oswaldo Cruz Institute for, 10, 157–159
Scientific institutions: role of in developing countries, 105; role of research in survival of institutions, 152
Scientific Society of Rio de Janeiro, *see* Sociedade Scientífica do Rio de Janeiro
Seabra, J. J., 87
Sellow, Friedrich, 26, 27
Serum Therapy Institute of Manguinhos, 9; founded, 68; Baron Pedro Affonso appointed director, 68; history 1900–1907, 68–69, 73–75, 95–96; Cruz appointed to staff, 69; resignation of Baron Pedro Affonso, 75; Cruz appointed director, 75; and the Pasteur Institute, 75–79; as a "crisis" institution, 79; effect of sanitation campaign, 91–98; congressional debate 1903, 92–95; new buildings begun, 96; renamed Institute of Experimental Pathology, 97; as strategic arm of yellow fever campaign, 107. *See* Oswaldo Cruz Institute
Sigaud, Joseph F.X., 50, 51
Sigüenza y Góngora, Carlos de, 17
Silva Lima, José Francisco, 54
Simond, P.L., 66
Smallpox in Rio de Janeiro, 90
Sociedade da Medicina do Rio de Janeiro, 24, 25; founded, 51; and epidemic disease, 52–53
Sociedade Litterária do Rio de Janeiro, 23
Sociedade Scientífica do Rio de Janeiro, 23
Sociedade Velosiana de Ciéncas Naturais, 27
Society of Medicine of Rio de Janeiro, *see* Sociedade da Medicina do Rio de Janeiro
Sousa, A.F. de Paulo: career of, 137
Spanish American science: in sixteenth and seventeenth centuries, 16–20
Spix, Johann B. von, 26
Spruce, Richard, 27
Stein, Barbara, 20
Stein, Stanley, 20
St. Hilaire, Auguste de, 26

Technology: nature of technology, 166; relation to science, 166–167
Technology in developing countries: limits to importation, 168; dependency theorists and, 167–169
Thayer expedition to Brazil, 27
Toledo, Maria de, 107
Toledo, José Martins Bonilha de, 147
Trypanosomiasis americana, 6; discovery by Chagas, 6, 118–119. *See* Chagas, Carlos, *and* Oswaldo Cruz Institute

Universities: colonial, in Spanish America, 16–18; University of Coimbra, 22; only founded in Brazil in twen-

tieth century, 127; as locus of research in industrialized world, 177–178; science in Latin American universities, 178
United States: colonial science in, 16
United States Geological Survey, 34
Unna, Paul Gerson, 140

Vargas, Getúlio, 128
Vasconcellos, Henrique de Figueiredo, 73, 116; joins Oswaldo Cruz Institute, 107; as staff member of Institute, 109
Veloso, Frei José Mariano da Conceição, 23, 27
Vianna, Gaspar de Oliveira, 111, 119

Wallace, Alfred R., 27
Weinberg, Alvin M., 175; criteria for scientific choice, 172–173
Wucherer, Otto: work on filariasis, 54

Yellow fever: investigated by João Batista de Lacerda, 33; epidemics in Brazil, 48; debate over causes, 48–49; lack of immunity in foreigners, 49; Pereira Barreto on, 58; returns to Brazil in 1849, 59; epidemic of 1878 in United States, 59; search for causative organism in South America, 59; work of Reed Commission, 65, 143; campaign in Rio de Janeiro, 87–91; Ribas' and Lutz' investigations, 142; in Campinas, São Paulo, 143; Reed Commission mentioned, 143
Yellow fever campaign in Rio de Janeiro: history of, 88–91
Yersin, Alexandre, 65–66, 72, 77